OPULENT OCEANS

Sterling Signature
NEW YORK

An Imprint of Sterling Publishing
387 Park Avenue South
New York, NY 10016

STERLING SIGNATURE and the distinctive Sterling Signature logo are registered trademarks of Sterling Publishing Co., Inc.

ISBN 978-1-4549-1341-2

Distributed in Canada by Sterling Publishing
c/o Canadian Manda Group, 165 Dufferin Street
Toronto, Ontario, Canada M6K 3H6
Distributed in the United Kingdom by GMC Distribution Services
Castle Place, 166 High Street, Lewes, East Sussex, England BN7 1XU
Distributed in Australia by Capricorn Link (Australia) Pty. Ltd.
P.O. Box 704, Windsor, NSW 2756, Australia

For information about custom editions, special sales, and premium and corporate purchases, please contact Sterling Special Sales at 800-805-5489 or specialsales@sterlingpublishing.com.

Manufactured in China

1 2 3 4 5 6 7 8 9 10

www.sterlingpublishing.com

The green pomacanthid angelfish on the back box cover taken from Mark Catesby's *The natural history of Carolina, Florida, and the Bahama island.*

NATURAL HISTORIES

OPULENT OCEANS

Extraordinary Rare Book Selections from the American Museum of Natural History Library

by

MELANIE L. J. STIASSNY, PhD

CONTENTS

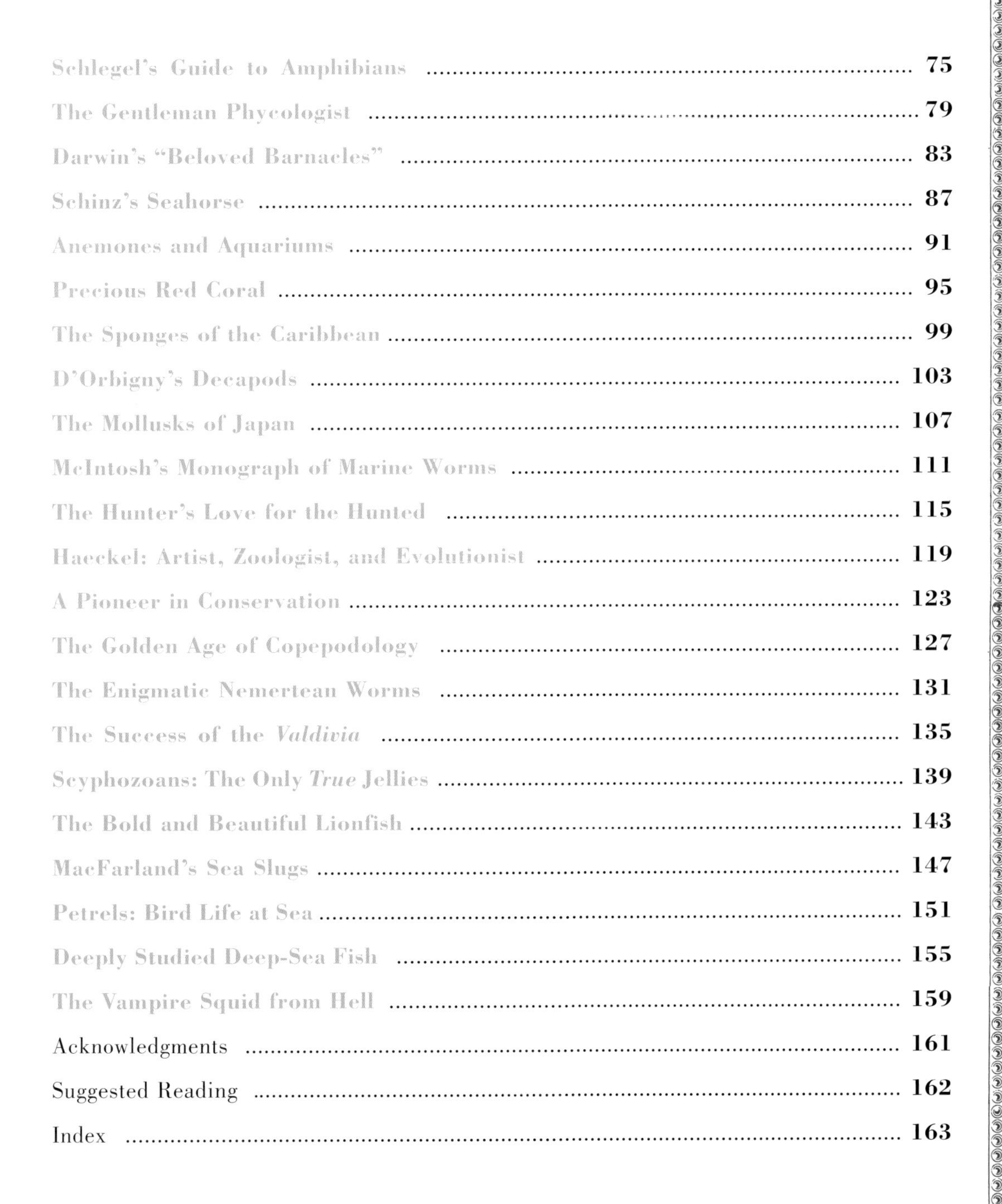

Foreword

ELLEN V. FUTTER

President of the American Museum of Natural History

Throughout human history, people have been linked with the ocean in the most fundamental ways. The basics of survival—food and oxygen—are provided by the sea, and civilizations have thrived alongside oceans, which offer myriad opportunities for transportation and trade. Beyond that, the human spirit is inherently drawn to the ocean for recreation, inspiration, and solace.

In this, the third compendium in the American Museum of Natural History's *Natural Histories* series, we are pleased to explore the world's "opulent oceans" and their beautiful and extraordinary inhabitants. This volume presents some of the most scientifically significant, historically rare, and beautiful illustrations of sea life from the Rare Books Collection of the American Museum of Natural History's Research Library, accompanied by illuminating essays by the Museum's Axelrod Research Curator, Melanie L. J. Stiassny, PhD.

The oceans are at the center of the American Museum of Natural History—quite literally. The magnificent Milstein Family Hall of Ocean Life sits at the heart of the institution, and is one of its most iconic and beloved galleries. Dominated by the ninety-four-foot blue whale model—representing the largest and most majestic creature ever to live on Earth—the Hall transports visitors to the glorious and sometimes strange underwater realm. Dazzling exhibits present the diverse ecosystems and species of the world's oceans. The Hall also is home to some of the Museum's most evocative habitat dioramas, including the two-level depiction of the Andros coral reef and cay above and below sea level, and the mysterious scene of a giant squid and whale locked in mortal combat.

The Milstein Hall, for which Dr. Stiassny was the lead curator, is the showcase for the Museum's long-standing commitment to marine research—of modern day fishes, invertebrates, and mammals; marine birds; marine paleontology; and, through the Museum's Center for Biodiversity and Conservation, conservation of marine ecosystems around the world. Though oceans make up two-thirds of the Earth's surface, and scientists throughout the ages have sought to study them, oceans are still among the greatest frontiers of science. Much of the oceans remains unexplored by science, even while habitats and species are under environmental threat.

Today, underwater submersibles, innovative diving equipment, and new imaging technologies are revolutionizing the way scientists, including those at the Museum, conduct research in the oceans, promising even more discoveries and scientific advances, like those depicted in this volume from earlier generations of naturalists.

We hope that this book will reveal and immerse you in the full richness of the oceans as captured by great scientists and artists throughout time and, in turn, will inspire and spark curiosity in you to learn about and seek to protect the oceans of today and tomorrow.

Preface

TOM BAIONE

Harold Boeschenstein Director of Library Services of the American Museum of Natural History

When the American Museum of Natural History's Library first began collecting nearly 150 years ago, our mission was to create a repository of recorded scientific thought and observation throughout the centuries. Although much has changed since 1869, the Library's original commitment has not. What has changed is our perspective of Earth's great blue frontier. As you will see on the following pages, early scientific exploration of the ocean was carried out by naturalists, explorers, and adventurers who endured storms and sickness while living in less than comfortable quarters. These discovery-seeking scientists' risk-taking was rewarded with a bounty of discoveries. Their efforts—exploring new areas and identifying new organisms—helped compile the many puzzle pieces of scientific data that continues to be fit together today.

Likewise, the book publishing landscape today is very different than it was a hundred or more years ago, when many of the treasures described in *Opulent Oceans* were created. While few would describe twenty-first-century publishing as simple, it is inarguably more straightforward and streamlined when compared to getting a book to print in the seventeenth century. The art of marrying ideas and images to print was an effort that often brought both financial and personal ruin to many—in some cases publish *and* perish was the norm.

The rich collections amassed by our predecessors in the Library—a small selection of which you will learn about in the following pages—present a broad and deep selection of riches one would need a lifetime to fully appreciate. The following pages document the histories behind the origins of many of these remarkable works of science and art, and will keep readers rapt with tales of the new and often strange creatures encountered.

In many ways, a great library has much in common with oceans: it is deep, rich, mysterious, and still retains some unexplored regions. As we begin to appreciate the process of creating these great books, we cannot help but be awed by the passion and creativity of their makers. We continue to find new uses for the information in these volumes, as they contain critical documentation and accumulated knowledge about our planet and its seas in the past, which is used today to help protect the very creatures that populate their pages. The extraordinary images prove that documenting species can be both aesthetically pleasing as well as scientifically informative, and our hope is that this book provides a fresh take on these illustrations, proving that these "old tomes" have a great deal of life in them yet.

Introduction

MELANIE L. J. STIASSNY, PhD
Axelrod Research Curator, Dapartment of Ichthyology, Division of Vertebrate Zoology

On December 7, 1972, astronauts on the Apollo 17 mission to the moon took one of the most celebrated images in human history—the astounding "blue marble" image of our planet as viewed from space. Suddenly, and for the first time, we saw not only how fragile our planet seems, how isolated in the vast expanse of space, but also just how little of it consists of actual earth. For we now know that ours is truly the "blue planet," with over 70 percent of its surface covered by a complex and interconnected three-dimensional world of saltwater—the World Ocean. Our understanding of the importance of that ocean to all life on the planet has been slow in coming and has over the millennia only gradually rendered up its secrets to us, an essentially terrestrial species.

The history of discovery in the ocean is long and storied. Yet it is salutary to realize that many of these discoveries, which have so fundamentally changed our way of thinking about the planet, have actually been made within the span of just the last fifty years. These findings are as profound as how the ocean makes the earth habitable, produces most of the atmosphere's oxygen, cycles carbon and nitrogen, and distributes heat across the entire surface of the planet. Deep seafloor cores reveal in minute detail the long history of the earth's climate. And changing ocean chemistry now warns of how activities on land are impacting climate today. Entire ocean biomes, tremendously rich in life—such as massive deep-water coral reefs, or the ubiquitous sea mounts that dot the ocean depths and aggregate organisms around their slopes—were, until surprisingly recently, virtually unknown. And a mere forty years ago, the stunning discovery of thriving biological communities living in total darkness around deep-sea hydrothermal vents—supported by microorganisms capable of converting searing hot minerals from the earth's fractured mantle into organic matter (chemosynthesis)—became part of an ongoing revolution in our understanding of life on our planet. Far from being a monotonous blue expanse mostly devoid of life, we now know that the ocean contains about three hundred times the habitable volume that land provides and is home to a truly astonishing diversity of life forms, from the majestic blue whales—the largest animals ever to have lived—to the recently discovered teeming microbial life so abundant yet so small that twenty thousand different kinds of marine microbes have been found in a single liter of seawater.

About two thousand years ago, the Roman chronicler Pliny the Elder confidently listed a grand total of 176 species of animals occurring in all the seas. And in 2010, the first ever Census of Marine Life—a global network of marine researchers participating in a ten-year study to document life in the oceans—estimated that there are likely three times as many unknown (unnamed) marine organisms as known, and that the grand total for marine life could easily top one million species. We really are still in the midst of the great age of ocean discovery and, as an active researcher in

1. A "fish eye" view of the Milstein Family Hall of Ocean Life at the American Museum of Natural History featuring a 29-meter (94-foot) long blue whale model.

this field, I have found it truly inspirational to delve into the past of that discipline and to look back on the lives and contributions of some of those who helped lay the foundations of ocean science and bravely led the way on this continuing journey of discovery of life in our planet's oceans.

For most of the essays in this book, I feature many well-known (along with a few unsung) heroes of what must have been an extraordinarily challenging and unbelievably exciting time of ocean exploration. The years spanning the eighteenth and nineteenth centuries were a time of the great voyages of discovery into uncharted regions and biomes that both metaphorically and literally opened up new horizons of geography and of knowledge of ocean life. For the naturalists who accompanied these voyages, and for those charged with identifying and describing the hundreds of thousands of specimens that were flooding into European and American museums from all around the globe, I can only begin to imagine the exhilaration. This, too, was a golden age for scientific illustration, and along with many talented amateurs, some of the world's leading artists and engravers were recruited to render the stunning new discoveries. The scientific reports produced during this period are replete with spectacular illustrations of innumerable new marine species. And it is these beautiful works that are recorded in the superlative Rare Book Collection of the American Museum of Natural History, some of which are featured here. A few of the artists whose work is found in the following pages were the wives or daughters of the scientists, yet, despite obvious talents, none were acknowledged as scientists themselves. I tried very hard to find women whose works could be highlighted in this book, but, in truth, a deep and pervasive hostility to the participation of women in science (particularly field science) tenaciously persisted up until the late twentieth century. Sadly, I could find not a single volume in the Museum's Rare Book Collection containing the work of a female marine naturalist over the entire time span covered in this book.

By the beginning of the twentieth century, photography increasingly began to replace the meticulously done, often sumptuously colored, engravings as the method of choice to illustrate new marine species. And while discovery rates continued unabated through that century and into our own, the depictions produced, although accurate, began to lose the artistic appeal so magnificently evident in the plates I have selected to accompany this volume. It is for this reason that my selection of essays covers up to the early years of the twentieth century.

But, of course, discovery continues and, while much has changed in the years that followed—happily including the burgeoning participation of women in the field—it is satisfying to recognize that many of those changes are simply of the technological accoutrements of modern science. Our fundamental mission to discover, describe, and understand the complex interrelationships of life on Earth, and the excitement that that engenders, remains for us today very much the same as it was for the early marine pioneers whose stories fill the pages that follow.

Book Conservation

BARBARA RHODES

Conservation Manager at the American Museum of Natural History Library

If one thinks of books, such as those described here in *Opulent Oceans,* as a resource, like the oceans themselves, it could be said that she or he is a bit of a romantic, but not necessarily wrong. Books and documents embody and contain the vast sea of human knowledge formed over the millennia since writing and printing were first invented. As with conservation of Earth's bodies of water, when the survival of part of this body of human knowledge is threatened, it sometimes requires specialized knowledge, skills, and training to pull it back from the brink.

On a more practical level, books are composites of materials, which include paper, inks and dyes, thread, adhesives, and various types of covering materials, such as leather or cloth. It is the nature of all of these materials to deteriorate, and that deterioration is irreversible. And, although our collections of books and documents exist to be used, wear and damage are inevitable. Custodians of these historic book collections must find the balance between access to information and the long-term survival of their materials. As Paul Banks, one of the founders of the profession of book and paper conservation, stated, "no one can have access to a document that no longer exists," in his essay "10 Laws of Conservation."

It is a conservator's goal to preserve not only the intellectual content of books and documents, but also the original physical forms in which the content is presented as much as possible, since the authenticity of these originals, once lost, cannot be restored. Although we may reproduce a text, electronically or otherwise, no reproduction can contain *all* the information that is present in the original; however, it must be noted that reproduction is preferable to losing access to the information altogether.

Modern conservation treatments tend to be limited to actions that will make a book stable enough to allow for safe handling and study but also "retreatable"—even though no conservation treatment can be considered entirely reversible. That is, all physical treatment of library materials involves changing them to some extent. Where physical treatment is necessary, it should lean toward the least intrusive sorts of repair so as to retain as much of the book's original structures as possible. A "full treatment" conservation of a book may include disassembly, surface cleaning, washing and/or deacidification of the paper; repair and resewing of the text; and rebinding, either in its original cover or in a new one. This, and all repairs, should use durable and chemically stable materials in addition to permanently retaining any original components that have been removed. Throughout the treatment, everything should be documented with written records and photography.

Unfortunately, while conservation of individual items in library collections is important and most libraries would like to maintain all of their collections in good condition, individual treatments can be time consuming and expensive. Therefore,

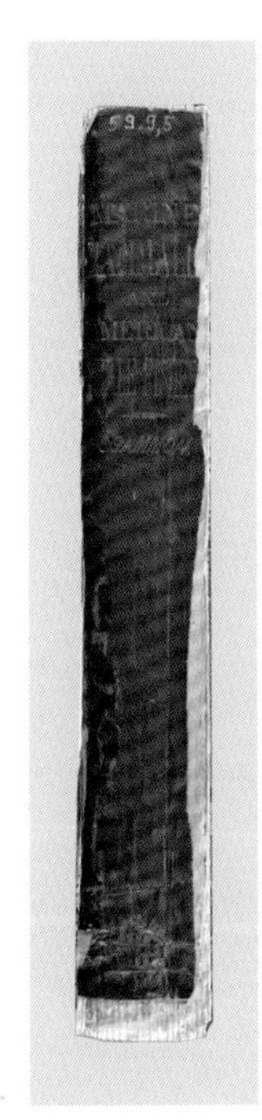

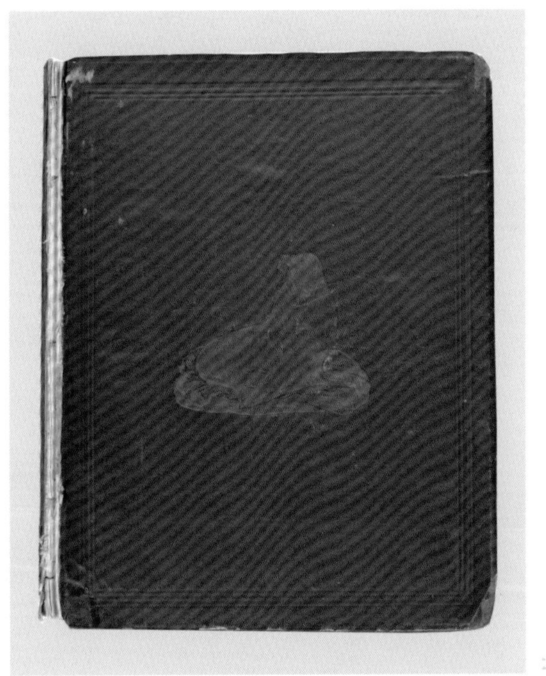

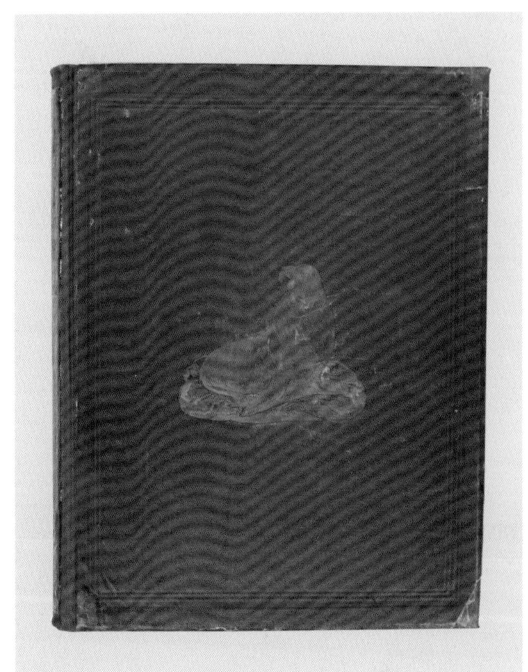

1. The spine of the cover of Charles Scammon's *Marine mammals*, which had become completely detached after being poorly repaired.

2. The front cover of *Marine mammals*, also detached from the book; the bookcloth had become cockled, and the corners were badly damaged.

3. The volume was given a new spine and board attachment using dyed cotton cloth; the corners were rebuilt and re-covered as well.

it is more cost effective for many libraries to place an emphasis on the *prevention* of damage to their collections by (1) maintaining proper storage conditions (the professional consensus recommends a stable 20 degrees C [68 degrees F] and relative humidity between 40 and 45 percent); (2) preventing infestation by mold and insects; (3) preparing for emergencies; (4) encouraging safe handling by staff and users through training and example; and (5) providing protective enclosures for vulnerable materials.

In many cases, enclosure may substitute for actual repair, especially for low-use items. Appropriate enclosures are determined for the book, pamphlet, or documents being conserved, which can be custom made or commercial boxes, envelopes, or wrappers. This not only allows the enclosed items to survive without alteration, but also keeps loose components together and protects them from handling and from environmental stresses. Enclosures may also be made for items that have been treated to protect them from further damage.

As a final thought, we turn to another of Banks's laws of conservation, which states that "conservation treatment is interpretation." A conservator cannot restore a damaged book to its original condition, but can often bring it back to a state that is both functional and aesthetically pleasing, which is important to how the book will be perceived and handled by the user. The conservator's knowledge of book structures and binding history is paramount to produce a result that is sympathetic to the book's original appearance and that allows the book to remain usable for an indefinite—but, hopefully, long—period of time.

.Lat. Capodoglio.It. Balene.

Debunking the Dolphin Myth

Author
Pierre Belon (1517–1564)

Title
La nature & diversité des poissons, avec leurs pourtraicts, representez au plus près du naturel

(Translation of *De aquatilibus, libro duo cum conibus ad viuam ipsorum effigiem, quoad eius fieri potuit, expressis*)

(*The nature and diversity of fishes, with their portraits represented close to nature*)

Imprint
Paris: Charles Estienne, 1555

1. Typical for his time, Belon's concept of "fishes" included marine mammals, and this simple woodcut depicts a charming, if rather fanciful, member of the dolphin Family Delphinidae.

The explorer, writer, and naturalist Pierre Belon lived a short but extraordinarily eventful life, rising from modest beginnings to become a favorite of two French kings and one of the most celebrated savants of the sixteenth century. Born near the Loire town of Foulletourte, he was apprenticed to an apothecary, and by the age of eighteen moved to the Auvergne as apothecary to the powerful bishop of Clermont. After travels through northern Europe, Belon gained the support of another powerful cleric, René du Bellay (1500–1546), scion of one of France's oldest families, who sent him to Wittenburg to study with the renowned physician Valerius Cordus (1515–1544). Belon accompanied Cordus on travels in the German provinces researching medicinal plants but in 1542 left for Paris to study medicine. Before obtaining a license to practice, he left to work for the powerful diplomat and military leader Cardinal François de Tournon (1489–1562). With de Tournon's support, in 1546 Belon set out on a voyage that would take him across the Ottoman Empire, from Greece to the Levant and beyond. He traveled for over three years, finally returning to France in 1549.

Still in the service of de Tournon, and while attending to the cardinal's household in Rome, Belon met another of de Tournon's protégés, Guillaume Rondelet (1507–1566), and the Italian Hippolito Salviani (see page 5). Both shared Belon's interest in natural history, particularly that of fishes, which at the time were understood to include all marine animals, from whales to invertebrates. Leaving Rome, Belon returned to Paris, where he continued preparing accounts of his travels and natural history observations. In 1551 his *Histoire de la nature des estranges poissons marins*, which featured the first images of sturgeons and tunas along with dolphins and other marine life, was published. Two years later the sensational *Les observations de plusieurs singularitez et choses memorables trouvées en Grèce, Asie, Judée, Egypte, Arabie et autres pays étrangèrs* was published to much acclaim. In that extraordinary compilation, Belon described numerous peoples and customs, ancient rituals and ruins, pharmacopoeias and medicinal cures, and countless animals and plants previously unknown to Europeans. His was the first "grand tour" that established the foundation and itinerary for scientific exploration for centuries to come.

Also published in 1553 was Belon's famous *De aquatilibus*, originally written in Latin but appearing in French translation in 1555 as *La nature & diversité des poissons, avec leurs pourtraicts, representez au plus près du naturel*. In that work, which is considered by many to represent the beginnings of modern ichthyology (the study of fishes), Belon described and illustrated over one hundred fishes, sharks, and rays. He also illustrated and discussed many marine mammals, including a number of dolphins, their embryos, and their reproductive anatomy. He considered dolphins "fish with lungs" and certainly not the monsters so often depicted on maps as smoke-billowing, fanged, or crested beasts threatening ships at sea.

Belon, unlike so many writers of his time, argued that we should rely on observation, and not be the "crooners of old songs, singing only out of habit, without ever really acquiring any musical understanding or skill." He made a clear association between marine mammals and land-dwelling animals, and his detailed descriptions and illustrations of their reproductive anatomy and embryos are considered to mark the beginnings of modern embryology.

The year 1555 also saw the publication of *L'histoire de la nature des oyseaux*, in which Belon famously illustrated the skeleton of a bird and a human side by side in upright posture with bones common to each similarly labeled—with that prescient juxtaposition, he suggested, for the first time, a unity of plan in the natural world.

In 1556 King Henry II honored Belon with a state pension and residence at the royal palace of Château de Madrid in the Bois de Boulogne and, after Henry's death, King Charles IX continued these privileges. Belon wrote on topics as diverse as agriculture and the desirability of acclimatization of certain exotic plants into France, to a treatise on ancient funeral rites and monuments. Sadly, at the age of just forty-seven and at the height of Belon's powers, unknown assailants killed this prescient observer of the natural world while he was crossing the Bois de Bologne.

2.

The figures accompanying *La nature & diversité des poissons* are simple hand-colored woodcut engravings and illustrate a range of marine organisms, but perhaps of greatest interest are the illustrations of dolphins, which, until Belon's groundbreaking studies, had often been inaccurately represented as fantastical sea monsters.

Dolphins and porpoises are marine mammals, which, together with whales, belong to the Order Cetacea (see page 117). There are thirty-two species of marine dolphins (including orcas and pilot "whales") in the Family Delphinidae and six species of porpoises in the Family Phocoenidae. Dolphins and porpoises are most easily distinguished by features of their snout and teeth, with dolphins tending to have prominent beaklike snouts bearing conical teeth and porpoises having shorter, blunter snouts and jaws armed with distinctive, spade-shaped teeth. Cetaceans, like other eutherian mammals, nourish their young in the uterus through a placenta, and feed them milk after birth. Gestation times range from about a year in the smaller species to around fifteen to eighteen months in the orca, which is the largest dolphin species. Dolphins commonly give birth to a single calf, which is born tail first. This is the case for all cetaceans, but is highly unusual for mammals, which generally give birth to their young headfirst. However, a tail-first birth makes perfect sense for a marine mammal that breathes air—a protracted headfirst birth would mean the calf would drown before its mother could push it to the surface to take its first breath.

2. More skeptical than many, Belon, nonetheless, included in his compendium some truly fantastical creatures such as this "sea-monk." Later, workers suggested that the creature, which appeared in numerous other compendia of the time, was possibly based on a stranded squid of some kind.

3. Despite its rather strangely webbed talons, Belon's charming young animal is likely a juvenile Mediterranean seal *Phoca vitulina*.

4. Although with an exaggeratedly horselike face and posture, Belon's drawing clearly depicts a seahorse, likely the common Mediterranean species *Hippocampus guttulatus*.

5. This illustration of the fabled web-footed horse of Neptune closely resembles images of the mythical creature, which were probably seen by Belon in Roman mosaics during his travels.

16 DE LA NATVRE ET POVRTRAICT

chassé aux riuaiges de la mer, tant pour l'usage, que lon prend de sa peau (que l'eau ne peult percer, & dit lon qu'elle garde du tonnerre) comme pour ce qu'il ha la chair de goust de sauuaigine.

Phoca. Gr. Vitulus marinus. Lat. Vecchio marino. Ital. Veau ou loup de mer.

3.

446 DE LA NATVRE ET POVRTRAICT

Hippocampus, Hippidium, Hippus en Grec & Latin, Caual marin en Francois, Falopa a Venise.

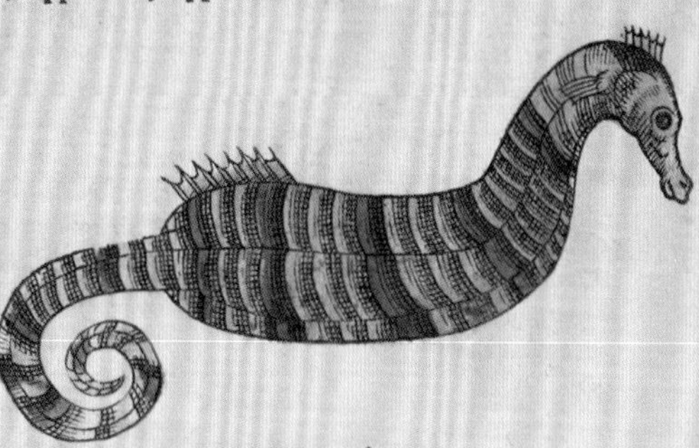

L'Orueul marin.

Le serpent qui estoit anciennement nommé Typhle ou Typhline, est maintenant ueu uulgaire es isles Ciclades, si semblable a un poisson, que le uulgaire Grec nõme Nerophidia, & a Marseille Gagnola, qu'on les prendroit l'un pour l'autre. Quand ceulx de Marseille peschent, & qu'ils ont apperceu un tel poisson en leurs rets, ils esperent auoir bien gaigné, & bon heur: c'est de la qu'il est ainsi nommé. Il est du tout inutile a manger. C'est un poisson de riuage, qu'on ne prend iamais a l'haim: sa bouche est si petite, qu'a pei-

4.

22 DE LA NATVRE ET POVRTRAICT

Le fabuleux Cheual de Neptune.

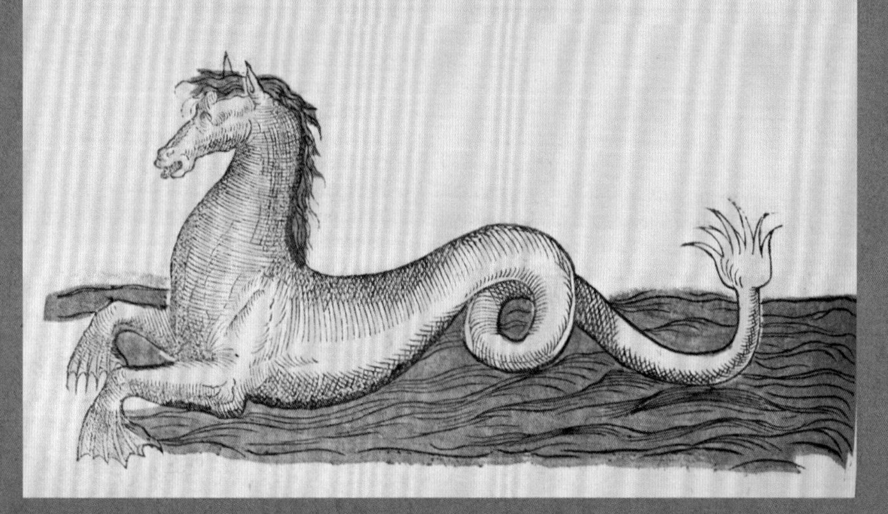

5.

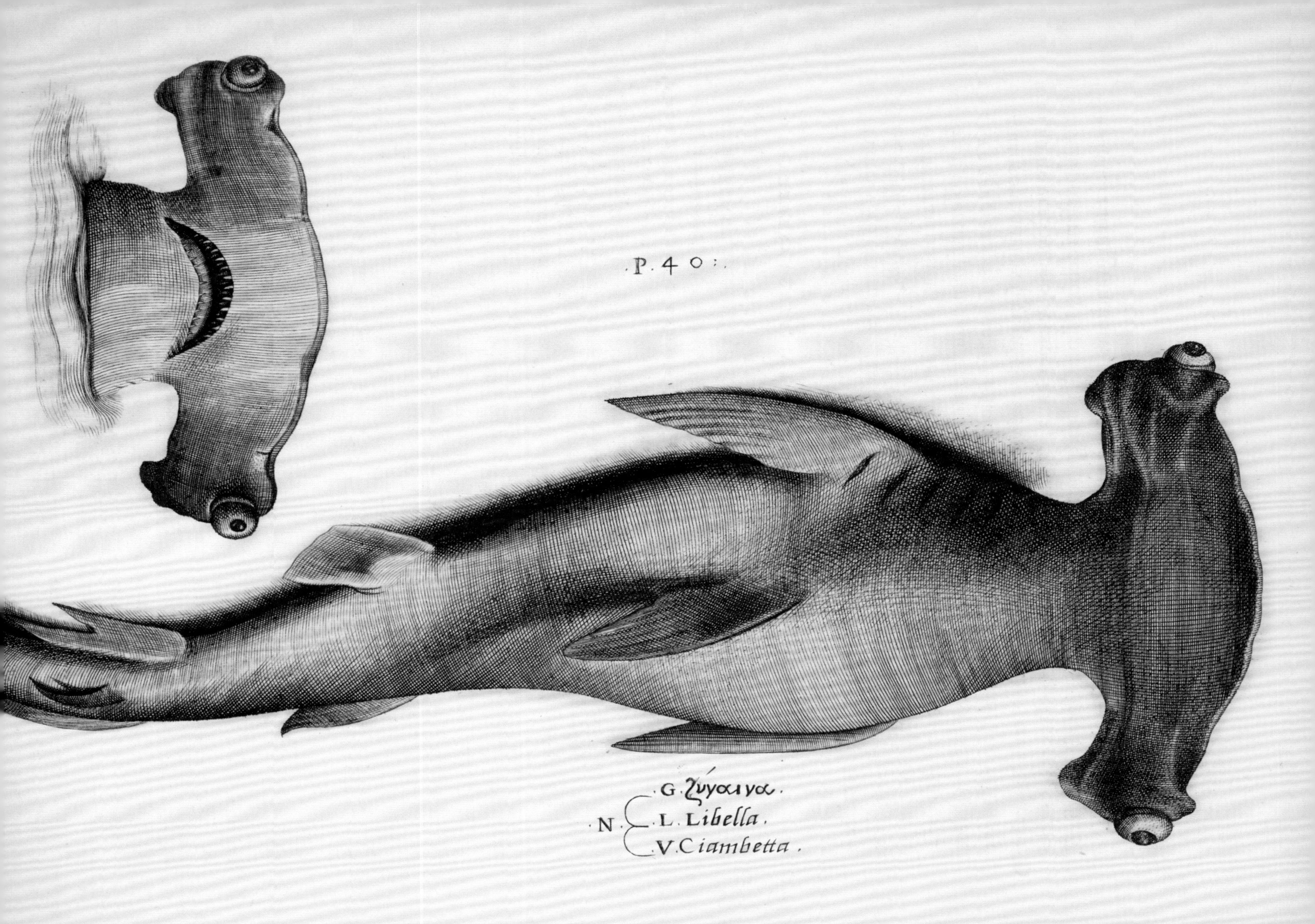
.P.40:.
.N. .G.Ζύγαινα.
.L.Libella.
.V.Ciambetta.

Mediterranean Mystery: Hammerhead Sharks

Author
Hippolito Salviani
(1514–1572)

Title
Aquatilium animalium historiae, liber primus, cum eorumdem formis, aere excusis.

(A natural history of aquatic animals. Book one, with engravings of their forms.)

Imprint
Romæ: Hippolyto Salviano, 1558

1. This beautiful copper plate engraving of a smooth hammerhead shark (*Sphyrna zygaena*) illustrates the surprisingly modern aspect of many of the plates in Salviani's groundbreaking work.

Hippolito Salviani was born in the northern Umbrian town of Città di Castello. After classical schooling in Umbria, he moved to Rome where he studied medicine and practiced as a physician. While in Rome he developed a keen interest in natural history, particularly that of fishes. Salviani's talents soon attracted the attention of Cardinal Cervini (1501–1555) (later to become Pope Marcellus II), who sponsored his ichthyological explorations along the Italian Mediterranean coast. With Cervini as his benefactor, Salviani gained close association with the Vatican, eventually becoming personal physician to three successive popes. His position as papal physician lent considerable social standing, and in addition to Vatican duties and a wealthy medical clientele, he was appointed principal physician of the medical college of the Sapienza University of Rome, where he taught until 1568. As a naturalist he was held in the highest esteem and "when anything curious in animated nature found its way to Rome, he was almost invariably and immediately apprised of it." In this way Salviani developed considerable personal knowledge of the organisms he wrote about, and rather than relying on the writings of "the ancients," he declared a determination "to state nothing, the truth of which we had not ascertained." Like his French contemporary Pierre Belon (see page 1), Salviani's insistence on observation over authority was a herald of the seventeenth century enlightenment, and Salviani and Belon are rightly considered two of the founding fathers of ichthyology.

In 1558 Salviani's *Aquatilium animalium historiae* was completed in two elegant folio volumes, and the work was exceptional in many respects. Rather than simply repeating or modifying the often fantastical assertions of Aristotle and Pliny, Salviani limited his observations to animals that he himself had examined, many of which he had gathered from fishermen and in local markets. His personal knowledge of his subjects is evident in the text provided for each species, which records not only external appearance but also habits, behaviors and reproduction, methods of capture, nutritional and medicinal usages, and often even cooking techniques. He was the first to narrow the concept of "fish" from a broad notion embracing all marine animals to instead include only the bony and cartilaginous fishes (although he did discuss a few cephalopod mollusks in the work). And, finally, the illustrations accompanying the work were truly remarkable for the time—elegant, large-format copper engravings rather than crude woodcuts as had been employed for depictions of fishes prior to Salviani's groundbreaking work.

The illustrations in the *Aquatilium animalium historiae* are particularly noteworthy and the eighty-eight full-page copper engravings that accompany Salviani's text artfully lend a lifelike and animated appearance to many of their subjects, features sorely lacking in the crude woodcuts employed by Salviani's contemporaries. However, it should be noted that despite the pleasing appearance and semblance of accuracy of the plates, a number of the species depicted lack important details of scale placement and number, or are so stylized as to be unassignable to known Mediterranean species. Many others, such as the wonderfully rendered John Dory (*Zeus faber*) are readily recognizable and the *Aquatilium* stands as a truly extraordinary early contribution to Mediterranean ichthyology.

The identities of Salviani's artists are not known with certainty, but it has been suggested that the famed student of Michelangelo, Nicholas Beatrizet (1520–1560), is responsible for the beautifully ornamented portrait of Salviani and possibly some of the illustrations. Others have been attributed to the master engraver Antoine Lafréry (1512–1577). Salviani printed the *Aquatilium* at his own expense and on his own press, and the grandeur of the work is testament to his wealth and high position in the Vatican establishment. He had intended to dedicate it to his benefactor, Cardinal Cervini, but Cervini died in May 1555, just twenty-eight days after his election as Pope Marcellus II, and in his place the *Aquatilium* honored his successor, instigator of the Roman Inquisition, Pope Paul IV. Salviani continued teaching and tending to his lucrative medical practice until his death in Rome at the age of fifty-eight.

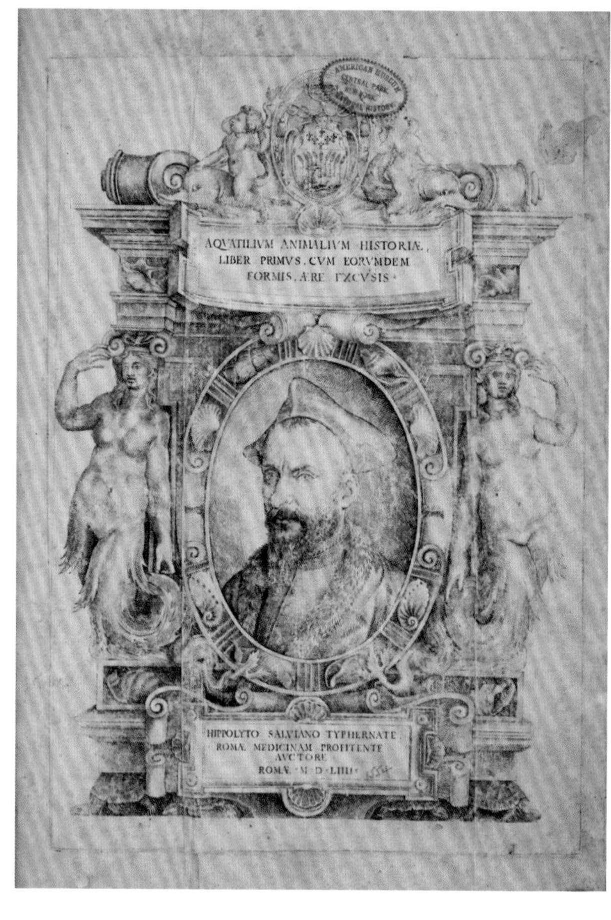

2.

Eleven species of hammerhead sharks are known worldwide, and all belong to the Family Sphyrnidae. Although they are closely related to other carcharhiniform sharks, such as the tiger and bull sharks, hammerheads are highly distinctive, with flattened, laterally expanded heads forming what is termed a cephalofoil. There is debate regarding the function of this strange structure, but recent studies have confirmed that lateral displacement of the shark's eyes at either end of the cephalofoil greatly enlarges their visual field, allowing them to see above and below at the same time. Additionally, all chondrichthyans (see page 68) have a well-developed "electric sense," which is able to detect very weak electric fields in water, and it has been suggested that the cephalofoil provides additional surface area for electroreceptors (ampullae of Lorenzini). Hammerheads have relatively small mouths and feed mainly on bottom-dwelling rays, fish, and crustaceans often buried in sand, and, since all animals produce an electric field when their muscles contract, it is likely that the cephalofoil may aid the hammerhead in electro-locating food that is hidden from sight or smell.

Of the three species of hammerhead sharks found in the Mediterranean (the great, the scalloped, and the smooth hammerhead), the species depicted by Salviani, with its smooth-edged cephalofoil and small second dorsal fin, is readily recognizable as the smooth hammerhead *Sphyrna zygaena*.

2. The lavish title page features a portrait of Salviani ringed by magnificent marine motifs and architectural pomp.

3. The exquisite rendering of this John Dory (*Zeus faber*) is an outstanding example of the high quality and lifelike appearance of many of the illustrations in Salviani's *Aquatilium*.

4. Readily recognizable, this goosefish (*Lophius piscatorius*) is, today, as it was in Salviani's time, a frequent addition to the famous Bouillabaisse soup of the Mediterranean.

5. Swordfish (*Xiphias gladius*) have been fished in the Mediterranean since ancient times, and in Salviani's day they commonly reached 3 meters (c. 10 feet) in length.

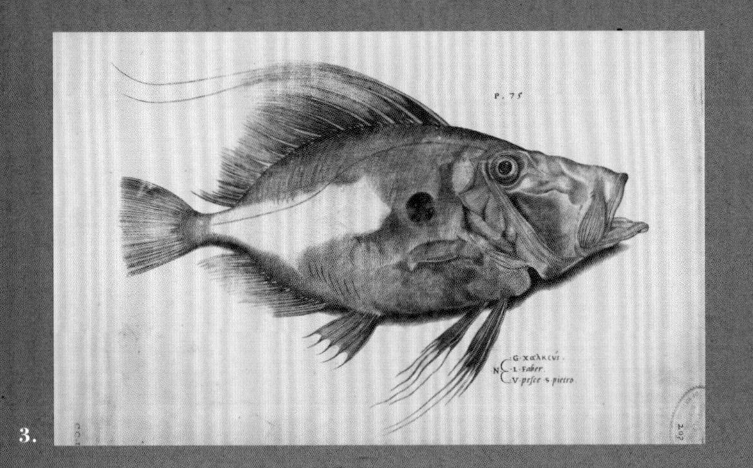
P. 75
G. χαλκεύς
N L. faber
V. pesce s. pietro

3.

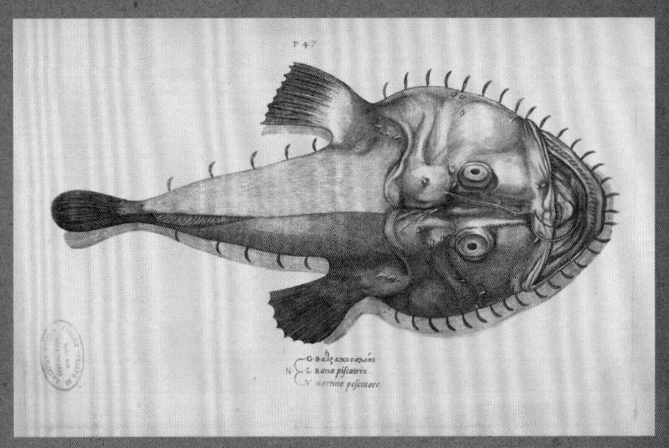
P 47
L. Rana piscatrix
V. marrone pescatore

4.

P. 39
G. ξιφίας
N L. Gladius
V. Pesce Spada

5.

A

Life of the Living Fossils

Author
Georg Eberhard Rumphius (1627–1702)

Title
D'Amboinsche rariteitkamer . . .

(The Ambonese cabinet of rarities . . .)

Imprint
Amsterdam: F. Halma, 1705

Figures 3 & 4
Author: Alphonse Milne-Edwards (1835–1900)
Title: *Recherches sur l'anatomie des Limules* (*Annales des sciences naturelles*, 5th ser., zool. t.17)
Imprint: Paris, 1873

1. Horseshoe crabs enter shallow water to mate, where males hold fast onto the females using a pair of modified claws. Rumphius was a keen observer, and his drawing clearly depicts a female with unmodified claws.

Georg Eberhard Rumphius was born and raised in Brandenburg-Prussia, where his father was an engineer in the town of Hanau. Rumphius's mother was the sister of the governor of the Dutch-speaking province of Cleves, and from a young age Rumphius appears to have been fluent in both German and Dutch. Little information on his early life is available, but we do know that at the age of twenty-four, shortly after the death of his mother, Rumphius entered the service of the Dutch East India Company in its military branch. He set sail for the Indonesian East Indies in late 1652 and arrived at the company's headquarters in Batavia (Jakarta) in July the following year. After a short stay in Batavia, he left for the Maluku archipelago (the Moluccas) and was based at the Dutch settlement of Ambon on Ambon Island (Amboina).

Ambon had been the headquarters of the Dutch East India Company from 1610 until 1619, when the company moved its center of operations to Batavia on Java. Nonetheless, the Dutch maintained a strong presence in the Moluccas and Rumphius rapidly rose through their ranks, attaining by 1657 the title of engineer and ensign. By this time he had embarked on his epic study of the flora, fauna, and ethnography of the region. His burgeoning interests led him to request transfer into the civilian ranks of the company, and in the same year he moved to Hitu on the north side of Ambon Island as a member of the merchant branch. Rumphius continued his naturalist studies in earnest, collecting and documenting everything, and it was during this time that he established a series of correspondences with scientific notables from across Europe. Although during his lifetime Rumphius published relatively few scholarly works, through his correspondences and conveyance of numerous specimens, his reputation was established. In recognition of the importance of Rumphius's work, the Batavia-based governor-general of the Dutch East Indies, Joan Maetsuycker (1606–1678), released him from ordinary company duties, allowing him to focus his attentions on scientific and ethnographic studies. Rumphius's primary interests were in the resplendent flora of the region, and his masterwork, which remains a comprehensive reference for the more than two thousand plant species of the region, is the magnificent *Herbarium Amboinense*, finally published in 1747, nearly forty years after his death.

Rumphius was fluent in Latin and Malay and, unusual for a man of his time, he befriended and worked closely with local Ambonese, who provided him with specimens and shared their considerable knowledge of the habits and uses of the local flora and fauna. By the late 1660s, his sight was failing and in 1670, probably due to glaucoma, he was blind, but even this could not halt his insatiable thirst for exploration. With the help of his beloved Ambonese wife, Susanna, and their eldest child, he continued collecting and annotating his works. At this point his writings turned from the customary Latin to Dutch, presumably because there was no one on the island who could transcribe Latin dictation. Seven years later an earthquake and tsunami struck the island and both Rumphius's wife and their eldest child were killed. Somehow he struggled on, only to be set back time and again by a series of catastrophes including the loss to fire of all the illustrations prepared for the *Herbarium*, and later the sinking of the ship carrying the finished manuscript and redrawn illustrations back to the Netherlands (happily one copy was retained on Ambon).

The single formal honor bestowed on Rumphius during his life was his election, in 1681, to the prestigious Germanic Academia Naturae Curiosorum, the first academy of sciences founded. At the time of his induction into that august society, Rumphius was given the honorific "Plinius" in recognition of his encyclopedic knowledge and contributions. And in the title page to his wonderful *D'Amboinsche rariteitkamer*, published in 1705, just three years after his death, he places the name "Plinius Indicus" (Pliny of the Indies) beneath his more formal title.

The *D'Amboinsche rariteitkamer*, describing hard and soft shellfish, rocks, minerals, and fossils, is a work of lesser scientific import than the monumental *Herbarium Amboinense*, but it is nonetheless of considerable interest and historical importance. It was published in three parts and beautifully illustrated with over sixty engravings. Rumphius's descriptions of the many marine organisms he encountered were detailed, for the most part accurate, and often included interesting observations on the uses of the animals he described. His beautiful rendering of an Indonesian horseshoe crab, for example, is accompanied by the following observations: "One finds it mostly in the near shore areas of Java, always on marsh flats and beaches, and always two by two, a husband and wife, and the wife must drag her husband, which is always smaller, on her back" and that "the Javanese will not eat it, saying it is harmful and produces dizziness and vertigo." Writing of the ubiquitous hermit crabs of the islands, he humorously notes that "these quarrelsome little creatures have caused me much grief, because when I laid out all kinds of handsome shells to bleach, even on a high bench, they knew how to climb up there at night, and carried off the beautiful shells, leaving me with their old coats."

It is hard not to admire Rumphius, who labored under a hot tropical sun in isolation from his homeland and fellow naturalists, faced innumerable losses and setbacks, yet soldiered on. In 1701 the surviving copy of his *Herbarium* reached the Netherlands, accompanied by a letter from the govenor-general of Ambon Island, who sadly recorded that "nothing more can be expected from the old gentleman, since he is finished with living." Rumphius may not have experienced adulation in life, but today he is recognized as one of the great tropical naturalists of the seventeenth century.

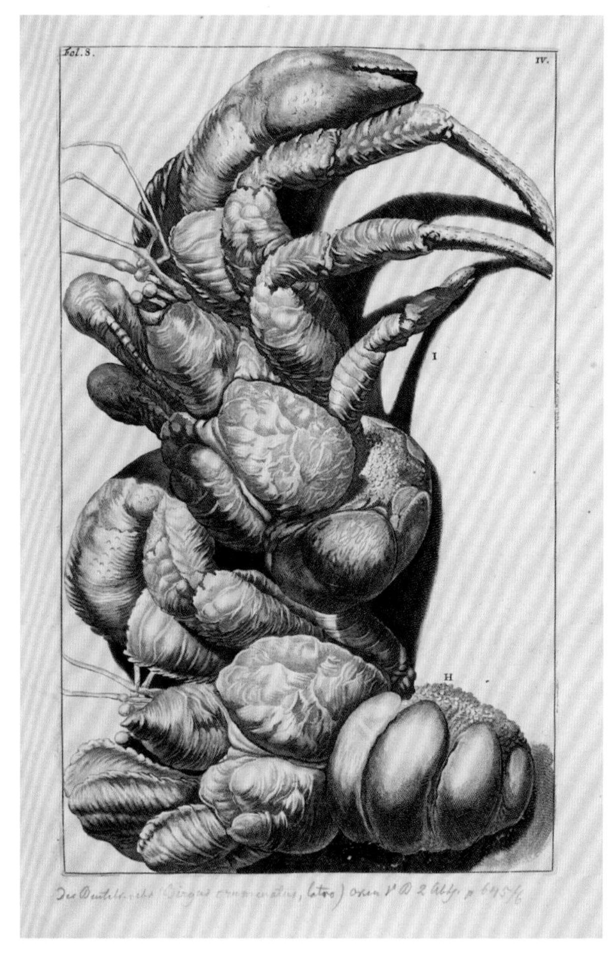

2.

2. The large size and strange appearance of the coconut crab (*Birgus latro*), along with its excellent culinary properties, made these crustaceans a favorite of Rumphius.

3. & 4. Almost two hundred years after Rumphius, the French naturalist Alphonse Milne-Edwards published in 1873 a detailed study of the anatomy of the horseshoe crab. Here, in red, the intricate arterial system is displayed, and in blue, the venous system.

Horseshoe crabs (also known as king crabs) were long thought to be crustaceans related to other decapods (see page 104). Of course, naturalists realized that these were very "odd" crabs; Rumphius, for example, chose the name *Cancer perversus* for his Javan specimens. But it wasn't until 1881 when the celebrated British biologist Sir E. Ray Lankester (1847–1929) published definitive proof that they were not crustaceans but marine relatives of arachnids (spiders and scorpions). The four living species, one Atlantic and three Indo-Pacific, are grouped in the Family Limulidae, within the arthropod Order Xiphosura. Their fossil record dates back almost 450 million years, and the living forms differ little in external appearance from these fossils, so are sometimes called "living fossils."

Modern horseshoe crabs spend most of their time on the seafloor feeding mainly on worms and mollusks. They only come ashore to breed, with each female laying between 60 and 120,000 eggs. Numerous seabirds and fish prey upon their eggs, and in this way horseshoe crabs play an important role in the in-shore food web.

In 1956 it was discovered that cells (amebocytes) in the blue, hemocyanin-rich blood of horseshoe crabs coagulate in the presence of bacteria or the endotoxins produced by them. Today Limulus Amebocyte Lysate is widely used by the pharmaceutical and medical industries to test for contamination of products. Horseshoe crabs can be bled and survive, but the large numbers taken from the wild for medical applications, combined with loss of coastal breeding habitat and the fishery for them as bait for conches, has resulted in a sharp decline in their numbers. Habitat conservation measures and restrictions on their fishery are necessary to ensure that these extraordinary "living fossils" continue to thrive.

3.

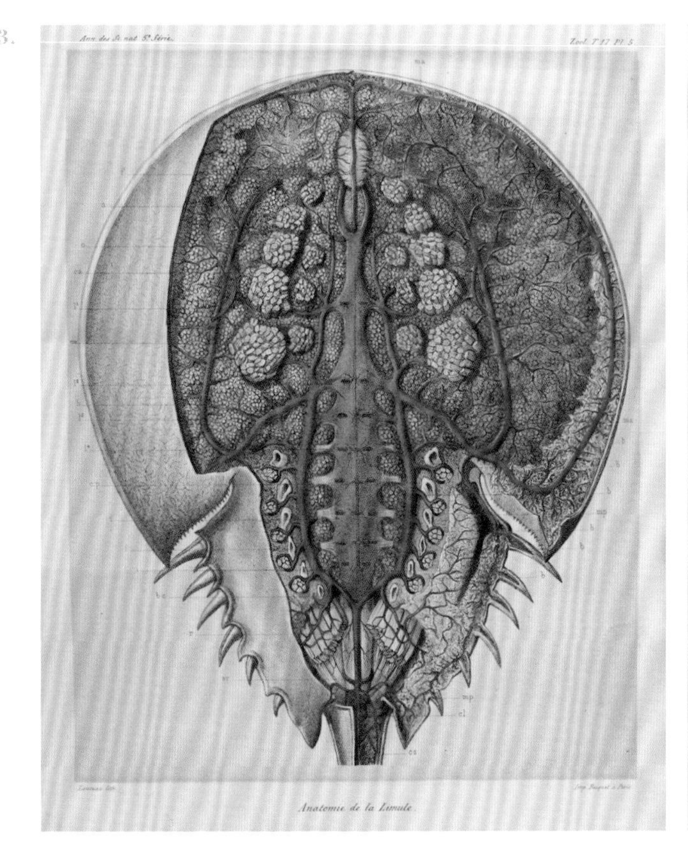

4.

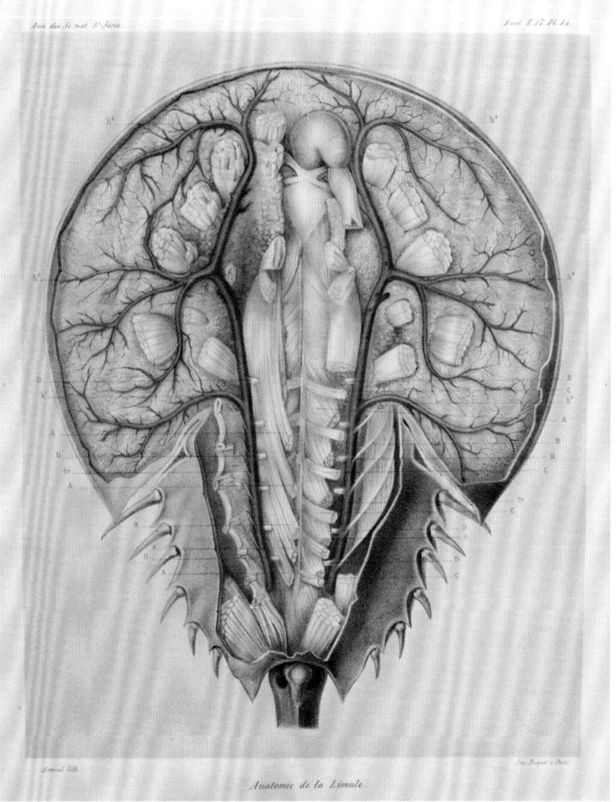

Tapia &c.

Catesby's Crabs

Author
Mark Catesby (1683–1749)

Title
The natural history of Carolina, Florida, and the Bahama island: containing the figures of birds, beasts, fishes, serpents, insects, and plants: particularly, the forest-trees, shrubs, and other plants, not hitherto described, or very incorrectly figured by authors

Imprint
London: Printed at the expence of the author, and sold by W. Innys ..., 1729–1747

The English explorer Mark Catesby was a pioneer naturalist whose beautifully illustrated account of life and nature in early Colonial America was among the first to highlight the natural wonders of that continent. His writings and collections of things "new and strange to the inhabitants of this side of the globe; and which must therefore necessarily excite" did indeed inspire a tremendous interest in the natural history of the Americas, and Catesby laid the foundation for many studies that followed.

Although details of Catesby's early life are scant, we do know he was born into comfortable circumstances in the town of Castle Hedingham, where his father was a successful lawyer. The family owned an estate close to the Essex border, and a neighbor and family friend was the preeminent naturalist and theologian John Ray (1627–1705), who likely stimulated Catesby's early enthusiasm for natural history. Catesby was twenty years old when his father died. After studying natural history in London, he left to visit his sister Elizabeth in Colonial Williamsburg. Elizabeth had married the prominent physician William Cocke (1672–1720) and the couple had left for America where Cocke was appointed secretary of state for the colony of Virginia. Catesby happily settled into Virginia high society, meeting others with interests in travel and natural history, most notably Colonel William Byrd II (1674–1774), an enthusiastic amateur naturalist and the wealthy founder of the city of Richmond. Catesby traveled widely throughout Virginia, making numerous collections, mainly of botanical specimens that he sent to friends and scientific acquaintances in England. In 1714 he collected botanical specimens and seeds in Jamaica before returning to Williamsburg, where he remained until departing for England in 1719.

One recipient of Catesby's seeds was the experimental horticulturist Thomas Fairchild (1667–1729), a member of the newly formed Society of Gardeners, and word of Catesby's work in the colonies spread rapidly through London's scientific milieu. Influential fellows of the Royal Society, led by a botanical colleague of Ray, William Sherard (1659–1728), urged the society to sponsor Catesby on a trip to the Carolina Lowcountry—at the time a virtually unknown and exotic wilderness. In 1722 Catesby returned to the Americas, where he spent four years exploring the Carolina and Georgia lowlands, then traveled south through Florida to the Bahamas. Everywhere he went he made watercolors and took copious notes on the bewildering array of plant and animal life he encountered, which were so unlike much he had known in England. Remarkably, he undertook most of this often grueling journey alone but for native guides, from whom he gathered information on the habits and uses of the specimens and nature he so keenly observed.

1. Catesby wrote that this colorful land crab (*Geocarcinus ruricola*) is so abundant when on breeding migration to the sea that "the Earth seems to move as they crawl about."

After Catesby's return to England in 1726, the ensuing seventeen years were spent working on his notes and drawings, which appeared in serial form and included over two hundred hand-colored folio-size copperplate etchings. These were ultimately compiled into the two magnificent volumes of *The natural history of Carolina, Florida, and the Bahama Island* published between 1731 and 1743. Catesby oversaw every aspect of production and engraved each copperplate himself. He drew the great majority of illustrations from paintings and notes he had taken from life but may have borrowed a few from other authors. A possible example is the beautiful Bahamian land crab (*Gecarcinus ruricola*) that accompanies this essay, as it has been noted that this crab bears striking resemblance, down to the precise position of its legs, to a figure drawn 150 years earlier by Sir Walter Raleigh's artist, John White (1540–1593). Borrowing images was not uncommon or frowned upon at this time, but what makes Catesby's image so interesting is the fact that in preparing the etching, he added the branch of a native Bahamian shrub, which the crab is grasping in its claws. In this way Catesby artfully augmented an otherwise sterile image with an allusion to the diet of the crab that, as we learn from his notes, includes the fruits of this exact plant.

Catesby's writings on the fauna and flora of the Americas were comprehensive, but his deepest passion was reserved for birds and botany. He had been elected a fellow of the Royal Society in 1733 and sixteen years later, in March 1749, he returned to a subject very close to his heart with a visionary paper on migration presented to the society and entitled "On birds of passage."

He died in his London home in December 1749, but fourteen years later his magnificent treatise on American botany, the *Hortus Britanno-Americanus*, was published posthumously.

Crabs (brachyurans), with nearly seven thousand species worldwide, are among the most highly modified of decapod crustaceans (see page 104). Although they exhibit a tremendous range of body forms, most are readily recognizable with their thick encasing exoskeleton and short abdomen typically entirely hidden under the thorax. Their size ranges from no bigger than a pea in some parasitic forms to giants such as some spider crabs with legs spanning 3.7 meters (c. 12 feet). Because of the way their legs articulate with the thorax, most crabs walk, or scuttle, sideways but some, such as the elegant *Portunus pelagicus* illustration accompanying page 132, have flattened paddlelike hind limbs and are extremely capable swimmers. Members of the crab Family Gecarcinidae are commonly called land crabs as they spend most of their adult lives on land, resting in underground burrows during the day to avoid dehydration, and foraging on fruits, vegetables, and insects at night.

Despite being well adapted to life on land, all must return to the sea to breed and lay their eggs. During annual breeding migrations, many thousands of crabs travel across land to reach the sea—a determined migration, which Catesby observed in the Bahamas and wryly noted "whatever they meet within their passage they go over, never going aside let houses, churches or what will stand in their way."

2. Catesby's Suillus is the gorgeous hogfish (*Lachnolaimus maximus*), a wrasse native to the western Atlantic Ocean and much prized for its flesh.

3. This moray eel (*Gymnothorax moringa*) is posed in a sea whip, and Catesby recounts that small fish seek refuge among its branches to avoid the voracious eel. The ecological juxtaposition of many of Catesby's compositions highlights an understanding of the natural history of his subjects.

4. Rather charmingly, Catesby notes of these crabs that structure is "so much better understood by the Figure of it than by the most tedious Description."

Suillus.
The Back Fin.

2.

Muræna maculata nigra
Lithophyton &c.

3.

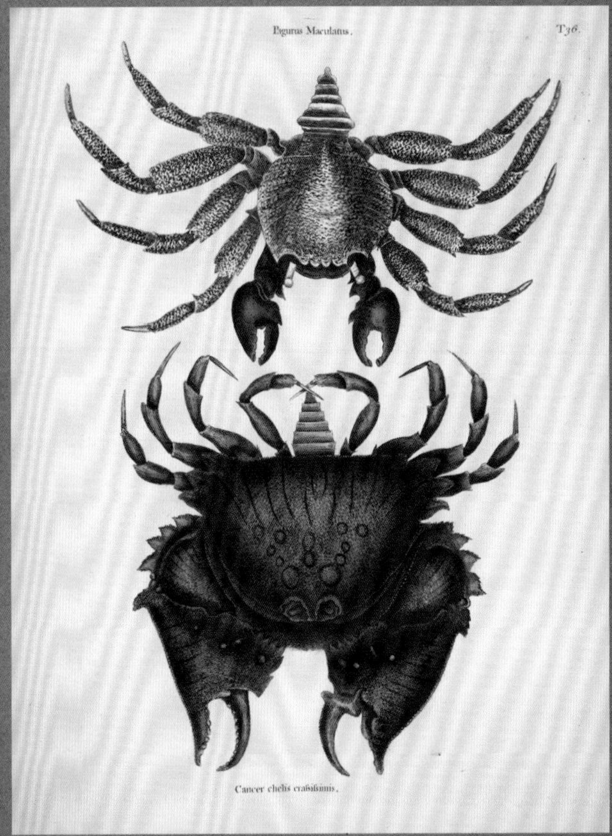
Pagurus Maculatus.
T36.
Cancer chelis crassissimis.

4.

Fig.I.
III.
IIII.

Discovering the Facts about Fossils

Author
Agostino Scilla (1639–1700)

Title
De corporibus marinis lapidescentibus quæ defossa reperiuntur

(On marine bodies which are found buried in stone)

Imprint
Romæ: Ex typographia linguarum orientalium Angeli Rotilii, et Philippi Bacchelli in ædibus maximorum, 1752

While perhaps better known for his artistic than for his scientific endeavors, the Italian Renaissance painter Agostino Scilla is nonetheless recognized as one of the founders of modern paleontology. Born in the Sicilian town of Messina in 1639, Scilla trained as a painter and apprenticed for five years with the baroque master Andrea Sacchi (1599–1661) in Rome, after which he returned to Sicily. Scilla was a successful painter of religious scenes and church frescoes, many of which are still in existence today. While in Messina, he was active in the Accademia della Fucina, a renowned Renaissance center of intellectual and political life where every Sunday, members would gather to debate ideas in literature, art, science, and poetry.

By the mid-1600s, Scilla had become increasingly interested in natural history, particularly in the abundant fossils he found in the Sicilian hinterland. He was struck by the strong resemblance of the petrified objects he found lying on the ground or embedded in rocks many miles from the sea, to living marine animals he was familiar with. At this time, naturalists throughout Europe were engaged in heated debate about the origin and nature of such fossils. The dominant theory expressed most influentially by the German Jesuit Athanasius Kircher (1602–1680) was that fossils were not the remains of formerly living creatures, but were formed in situ in rocks under the influence of a mysterious "lapidifying virtue" and therefore were not of organic origin. Scilla disagreed; as a keen observer, he noted the detailed similarities between the fossils he was finding and the seashells, urchins, and corals living in nearby waters. He wouldn't accept Kircher's unobservable lapidifying force and, because the similarities he documented were so striking, he argued that fossils were nothing other than the remains of living creatures that had turned to stone.

In 1670 Scilla presented his findings in *De corporibus marinis*, his only scientific publication, and in it he forcefully argued for the living origins of fossils. His philosophical approach, a clear precursor of skeptical empiricism, is most succinctly summarized in the volume's beautiful frontispiece: Entitled "Vanae speculationis sensus moderator" (Vain speculation undeceived by sense), Scilla depicts the figure of "Sense" demonstrating to "Vain Speculation" the clearly organic nature of the fossil sea urchins and shark teeth shown strewn on the hillside. Despite his conviction that fossils were once living organisms, he candidly admitted that he had no idea how these marine organisms could have arrived so far from the sea and couldn't think of a way of finding out. He did, however, speculate that they may have been deposited during the Noachian flood, or possibly in a series of floods.

Scilla was a keen observer of nature and his artistic abilities are evident in the many plates that accompany this volume. Most of the fossils illustrated are of whole

1. Sand dollars and sea biscuits are echinoderm relatives of sea urchins that tend to live buried in sand and sediment. They have a long fossil record dating back almost 60 million years.

2.

2. The beautiful, allegorical frontispiece to *De corporibus*.

3. Among the many fossils found in the Sicilian hinterland by Scilla were shark's teeth, like these beautiful specimens.

4. Scilla was particularly struck by the many similarities between the fossil sea urchins he collected far inland and those living in the sea nearby. It was this remarkable correspondence that led him to argue that fossils were once living organisms.

5. Here Scilla depicts a bluntnose sixgill shark (*Hexanchus griseus*), the species from which he posited some of the curiously shaped fossil teeth he found had come.

animals, mostly urchins and seashells, and there is also a selection of corals and fish bones; a few are of shark teeth. Fossil shark teeth are commonly found in European deposits but had not previously been recognized for what they were—instead, they had been considered magical objects called glossopetrae or tongue stones. Glossopetrae were believed to be either the tongues of snakes transformed into stone (or that they fell from the sky during lunar eclipses) and to have miraculous protective properties. Scilla was one of the first to correctly identify glossopetrae as shark teeth, even illustrating the species of sharks from which he thought they had come.

The Echinodermata is an exclusively marine phylum including many familiar animals such as sea urchins (see page 124) and starfish (see page 64), sea lilies and brittle stars (see page 26), as well as the less familiar sea cucumbers. Echinoderms have an unusual pentaradial body plan, although this distinctive fivefold organization is lost in adult sea cucumbers and a few other echinoderms. Sea urchins, sea biscuits, and sand dollars together form a highly successful group called echinoids. With almost a thousand species, echinoids are common from the intertidal zone to depths of over 5,000 meters (c. 16,404 feet). Although they come in many shapes and sizes, most echinoids have a hard outer skeleton or test made up of interlocking plates covered in spines. These spines may be long and prominent as in many urchins, or short and soft, forming a feltlike skin as in sea biscuits and sand dollars. In most echinoids, the mouth is on the undersurface and is ringed by five sharp teeth that form a complex chewing organ known as Aristotle's lantern. Echinoids feed mainly on marine algae but can also consume a range of invertebrates. In turn, they are the favorite prey of sea otters and many species of fish.

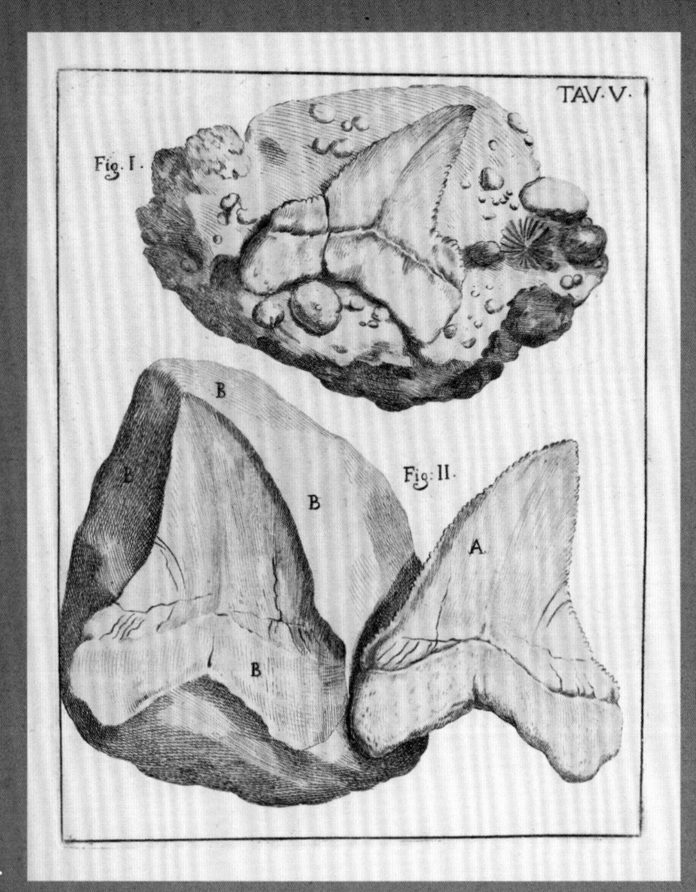
TAV·V.
Fig. I.
B
B
Fig. II.
A.
B
3.

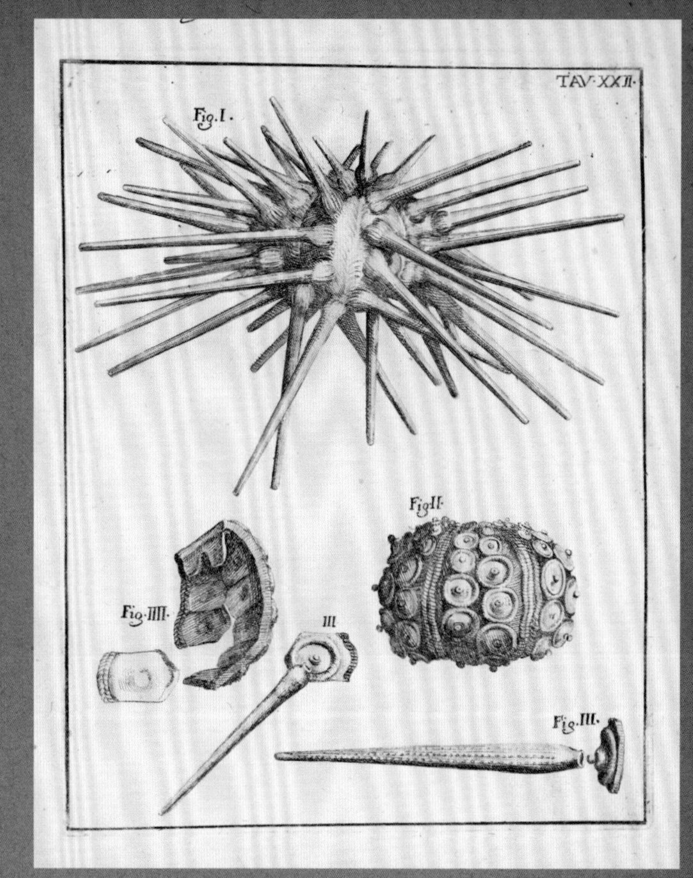
TAV·XXII.
Fig. I.
Fig. II.
Fig. IIII.
III
Fig. III.
4.

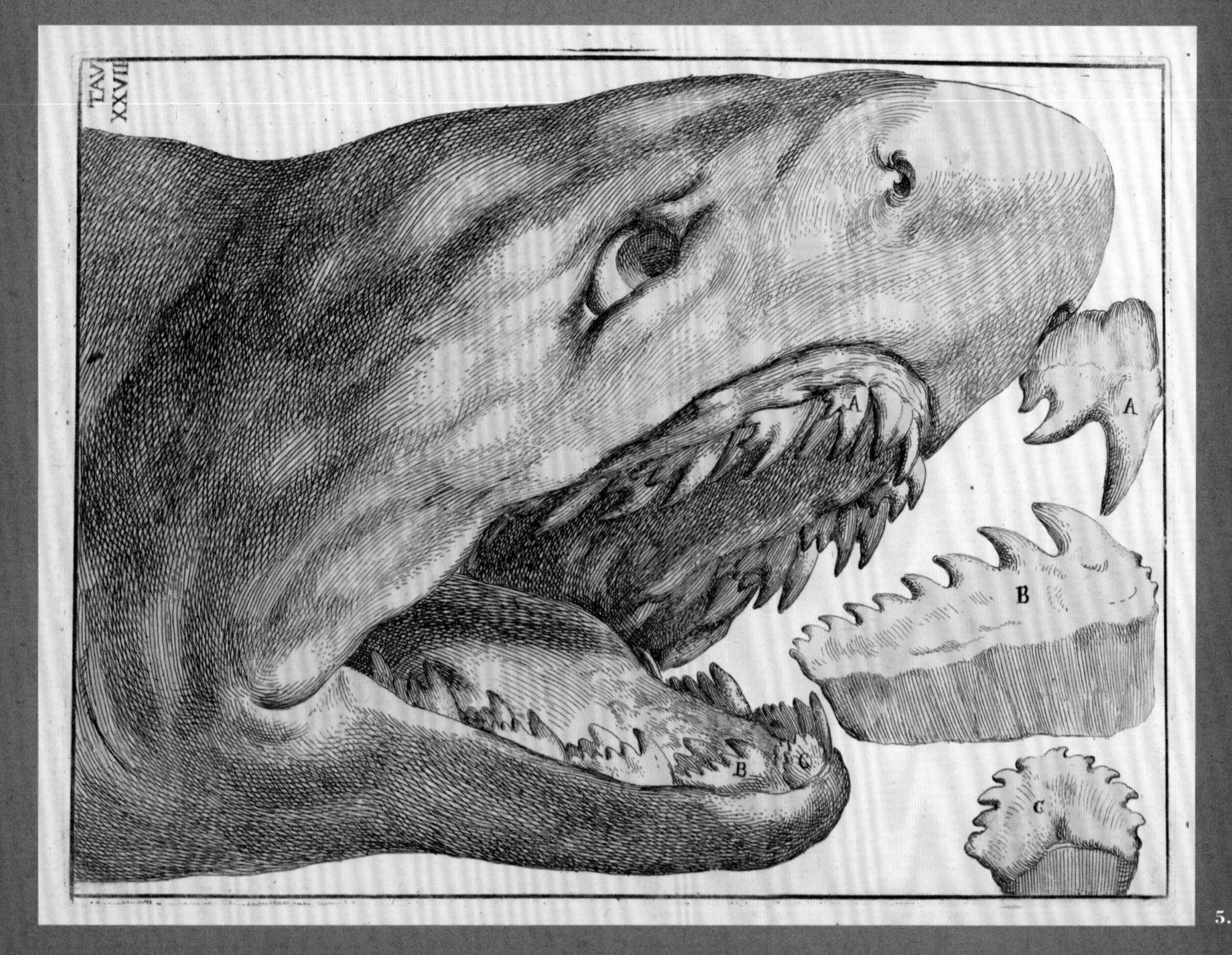
TAV XXVIII
A
A
B
B
C
5.

1.
2.

Knorr's Curious Cabinets and Corals

Illustrator
Georg Wolfgang Knorr
(1705–1761)

Title
Deliciae naturae selectae, oder auserlesenes Naturalien-Cabinet welches aus den drey Reichen der Natur zeiget, was von curiosen Liebhabern aufbehalten und gesammlet zu warden verdienet

(Selected natural delights, or a choice of everything that the three kingdoms contain worthy of the researches of a curious amateur to form a cabinet of selected natural curiosities)

Imprint
Nürnberg, 1768

1. Sea fans look like plants, but are animals related to corals and anemones. Their sturdy but flexible stems allow them to withstand strong ocean currents. Some can grow 2 meters (c. 6.5 feet) high and live hundreds of years.

The German illustrator Georg Wolfgang Knorr was a talented artist who became an influential art dealer and natural history publisher. He was born in the Bavarian town of Nuremberg, the son of a gifted artisan and woodworker and, after initially apprenticing in the wood trade, he soon turned to art and engraving. By the age of eighteen, Knorr was working for the Nuremberg publisher and engraver Martin Tyroff (1704–1758) alongside some of the most talented artists of the time on the extraordinary four-volume *Physica Sacra*. Written by the eminent Swiss physician Johannes Jacob Scheuchzer (1672–1733), the work blended biblical commentary with observations on the natural world and was accompanied by over seven hundred elaborate, often strangely fanciful, large copperplate compositions. Work on the *Physica Sacra*, which was published between 1731 and 1735, had ignited in Knorr an interest in the natural world, and in the following years, he gained a wide knowledge of art history and publishing as well as of the natural sciences.

During the second half of the eighteenth century, due in large part to the efforts of the distinguished physician and collector Cristoph Jakob Trew (1695–1769), Nuremberg had risen to prominence as a national center for the production of finely crafted, beautifully illustrated natural history books. Trew had gathered around him a select group of artists and naturalists, among whom Knorr was a prominent member, and together they contributed to many natural history publications. Around 1730 Knorr established his own publishing house, and in the following decades, he maintained a prodigious output of portraits, animal and plant studies, landscapes, and geological formations. He had become particularly interested in paleontology and between 1749 and 1755 published *Sammlung von Merckwüdigkeiten der Natur und Alterthümer des Erdbodens*. This publication was used years later by geologists in mapping stratigraphic successions based on many of the fossils and geological formations described and illustrated by Knorr. At about this time, Knorr began work on what was to become one of his most successful publications, the *Deliciae naturae selectae,* which was ultimately published posthumously between 1766 and 1768 by his heirs.

Knorr's is an iconic example of the natural history publications of the time, essentially an artistically rendered pictorial representation of the "cabinets of curiosities" so popular among the educated elites and royal houses then. Knorr drew and engraved many of the ninety-one magnificent, hand-colored copperplate illustrations depicting an interesting assortment of zoological and mineralogical subjects, and, although the accompanying text was anecdotal and far from scientifically rigorous, the illustrations were exceptionally vibrant and delicately rendered. Knorr himself drew many of the subjects depicted in the *Deliciae naturae selectae* from specimens housed in Trew's famous "cabinet" or from those living in Trew's private menagerie, and these are indicated as "Ex. Museo Excell. D. D. Chris. Jac. Trew" at the bottom of many of the plates. Other subjects were from notable cabinets of others in Knorr's acquaintance, and it is also probable that some were copied from engravings published by others.

Although Knorr does not provide reference to other works in the *Deliciae naturae selectae*, "borrowing" illustrations was not an uncommon practice at this time (see page 14), and it is certainly the case that many of Knorr's contributions to the botanical treatise *Thesaurus re herbariae hortensisque universalis*, published between 1750 and 1770, borrowed heavily without acknowledgment from the works of others.

Knorr died at the age of fifty-six in Nuremberg in 1761, some four years before the first folio volume of the *Deliciae naturae selectae* was published in an expanded form with a French translation and editorial additions by Philip Ludwig Statius Müller (1725–1776), professor of philosophy and natural history at the University of Erlangen. And in 1777 the popular work was published in Dutch translation.

The beautiful sea fan that accompanies this essay was probably added after Knorr's death and was drawn by his friend and fellow member of Trew's Nuremberg circle, the artist Christoph Nikolaus Kleeman (1737–1797), based on specimens held in the cabinets of Trew and of the work's editor, Müller.

DELICIÆ NATURÆ
SELECTÆ
oder
auserlesenes
Naturalien Cabinet
welches
aus den
drey Reichen der Natur
zeiget
was von
Curiosen Liebhabern
aufbehalten
und gesammlet zu werden verdienet.
Herausgegeben von
Georg Wolffgang Knorr
Nürnberg
1751.

2.

Sea fans and sea whips are colonial octocorals (see page 96), and about five hundred species are grouped together in the anthozoan Suborder Holaxonia. Unlike the hard calcareous skeletons of the stony corals (see page 52), the horny skeleton of most holaxonians is composed primarily of a complex protein (gorgonin) that, in sea fans, supports a delicate lattice of branchlets fanning out in a single plane. The sea fan's tiny coral polyps are embedded in the horny skeleton, and during feeding, each polyp extends eight tentacles to filter tiny zooplankton (see page 128) as they drift past in the current. The tissues of many holaxonians also house symbiotic photosynthetic algae (dinoflagellate species), which provide additional nutrients for their hosts.

Sea fans are found in tropical and subtropical oceans where, unlike many other octocorals and stony corals, they tend to anchor themselves in sand or mud, rather than to attach themselves to hard substrates. The size and shape of each individual is strongly influenced by location, and in shallow water where strong currents dominate, the fans are lower and wider than those in deeper, calmer waters. Sea fans are important components of coral reef ecosystems and provide shelter for many other organisms such as brittle stars and seahorses (see page 88), and are often encrusted by colonial hydrozoans, sponges, and other corals.

2. Knorr's beautiful frontispieces are renowned, and the *Deliciae* is no exception with its evocative scene of ocean bounty.

3. The veaux marin (*Phoca vitulina*) was said by Knorr to have the soft skin of a young cow.

4. Objects such as dried holaxonian sea whips and pieces of stony coral were popular items in collections of Knorr's time.

5. The much sought-after Head of Medusa basket star (*Gorgonocephalus caput-meduae*) was a particularly highly prized possession.

6. Dried outer tests of sea urchins, which retain their shape and color, were as attractive to eighteenth-century collectors as they are to seaside visitors today.

Ex Museo Excell. D.D. Christ. Iac. Trew.
Christian Leinberger ad nat. pinxit.
Andreas Hoffer sculpsit. 60

3.

A.VI
Ex Museo Excell D.D. Christ. Iac. Trew. f. f.
33.

4.

5.

D
Ex Museo D. Ioan. Ambrosii Baureri. Pharmacopoei Norimb
et Acad. Caesar. Leopoldino-Carol. Nat. Curiof. Socii celeberrimi.

6.

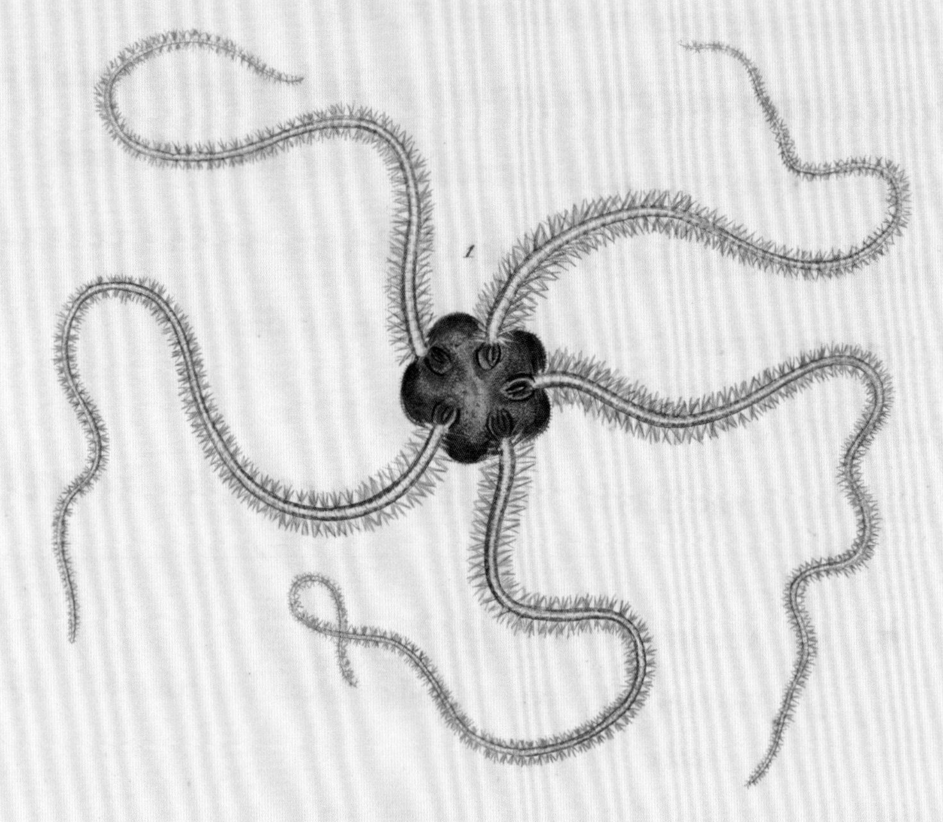

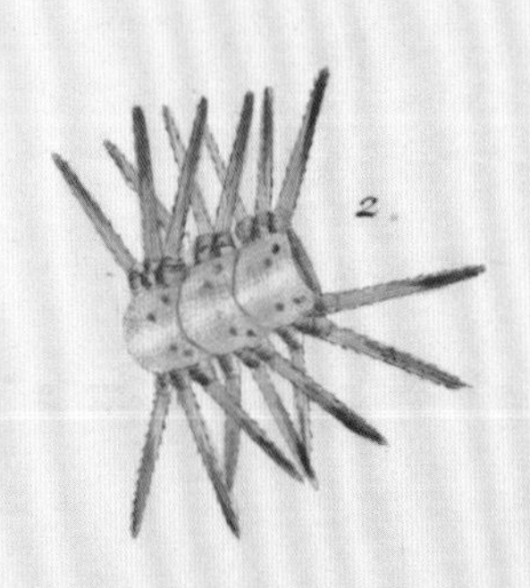

5

Zool. D. Tab. XCVII

Flexibility of Brittle Stars

Author
Otto Frederik Müller
(1730–1784)

Title
Zoologia danica, seu Animalium Daniae et Norvegiae rariorum ac minus notorum descriptiones et historia.

(Danish zoology, or description and history of the rare or less well-known animals of Denmark and Norway)

Imprint
Havniae: Typis N. Mölleri ..., 1788–1806

1. Brittle stars, like this colorful *Ophiactis tricolor,* are found in large numbers on the deep ocean floor, and in certain regions the ocean bottom can literally be swarming with thousands of individuals.

The pioneering Danish marine biologist and botanist Otto Frederik Müller rose from humble beginnings to become one of his nation's leading scientific figures, considered by the eminent Jean Léopold Nicolas Frédéric Cuvier (1769–1832)—commonly known simply as either George Cuvier or Baron Cuvier—as "in the first rank of those naturalists who have enriched science with original observations." Müller was born in Copenhagen, the son of a court musician who had little time or money to spare for his son's education. So at the age of twelve Müller was sent to stay with his uncle in Jutland on the North Sea coast. Under his uncle's direction, Müller studied history and music, and proved to be an excellent student.

Returning to Copenhagen, Müller enrolled at Copenhagen University, first studying theology and then law, but financial constraints forced him to leave in 1748 to earn his living as a musician. His fortune changed in 1753 when he was appointed private tutor to the son of the late Count Schulin of Frederiksdal (1694–1750). For the next sixteen years until the death of the Countess Schulin, Müller lived with the family in Copenhagen during the winters and at the family seat at Frederiksdal, on Lake Furesø, in the summers. Countess Schulin was an enthusiastic amateur naturalist, and it was she who ignited a similar interest in Müller, whose natural history studies began during his time in Frederiksdal.

Müller's first zoological work was on the insects of the region, and the *Fauna Insectorium Friedrichsdaliana* was published in 1764. He left soon after with the countess's son on an extended trip through central and southern Europe, where he met with many of the continent's scientific luminaries. Returning to Frederiksdal, he finished a local flora, published in 1767 as *Flora Freidrichsdaliana*. Müller's flora attracted the interest of King Frederick V, who commissioned him to work on the monumental *Flora of Denmark*, an ambitious project sponsored by the Danish royal family. Müller produced five fascicles for the multiauthored *Flora Danica*, a massive endeavor not fully completed until 1883.

When in 1769 his benefactor the Countess Schulin died, Müller was required to take a series of government posts and, although he continued to publish important scientific papers, administrative duties occupied much of his time. Happily the situation changed in 1773 when he married into the wealthy Paludan family and retired from government service to devote himself fully to his natural history studies. Through his marriage to Anna Catharina, Müller had become the owner of an estate on the shores of Kristianafjord (Oslofjord) near Drøbak in Norway, and from 1773 to 1778, much of his time was spent there examining dredges made of the rich sea life of the fjord, and making similar collections in Danish and Dutch coastal waters. During this time, Müller developed a keen interest in a wide range of marine invertebrates, most particularly in the bewildering array of microscopic organisms, known at the time as "infusoria." His contribution to their description and classification was widely acknowledged as among his most important work.

In 1776 Müller published the *Zoologiae danicae prodromus,* a preliminary summary of the marine life he had discovered, with a classification for the more than three thousand native species, and by 1777 he had begun his masterwork, the magnificent *Zoologia danica, seu Animalium Daniae et Norvegiae rariorum ac minus notorum descriptiones et historia.* It had been Müller's intention that the *Zoologia danica* would become as comprehensive a compendium of marine animal life as was the *Flora danica* for the plants of the region, but he was able to complete only two volumes before his death. However, the work was continued under the direction of Peter Christian Abildgaard (1740–1801) and published in four richly illustrated folio volumes between 1788 and 1806. In these volumes a wide array of marine life is described, much of it for the first time, and illustrated with 160 copperplate engravings. Many of the plates do not bear the name of an artist and it is assumed that Müller, himself a talented artist, drew these. Others were prepared by the well-known Danish artist and engraver I. G. Fridrich, or by Müller's younger brother, Christian Frederik Müller (1744–1814), who helped complete the final volume. Otto Müller died at the height of his prowess in Copenhagen in December 1784. In addition to the completion of the *Zoologia danica,* a number of other important works, most notably on microscopic organisms and pelagic crustaceans, were published posthumously.

Among the vast array of animals featured in the *Zoologia danica* are a large number of echinoderms (see page 124), particularly brittle stars such as the delicate, tricolor brittle star featured in the plate accompanying this essay. Brittle, or serpent, stars belong to the Class Ophiurodea, and of the nearly two thousand species living in the world's oceans, about twelve hundred are found only in deep water habitats at depths greater than 200 meters (c. 650 feet). It is likely that many of the brittle stars that Müller encountered were from the dredges he employed to sample in the deep, cold waters of the Kristianafjord.

Unlike sea stars, brittle stars have flexible, elongate legs supported internally by a series of calcium carbonate plates articulated via ball and socket joints and controlled by muscles, enabling them to move rapidly across the seafloor. Like many other echinoderms, the mouth (which serves also as an anus) is on the underside of a central disk and bears five sharp teeth (Aristotle's lantern) used to grind up their prey, which consists of organic particles scavenged from the seafloor as well as small crustaceans and worms. They in turn are prey for crabs, starfish, and even other brittle stars.

2. Although these tiny hydrozoans (*Clava multicornis*) grow only 25 millimeters (c. 1 inch) tall, Müller's illustration is highly accurate in all anatomical details.

3. Skeleton shrimps, like these *Caprella linearis*, are slender-bodied amphipod crustaceans. Females (above) retain fertilized eggs in brood pouches, and their young emerge as juvenile adults. Müller recognized the males (below), although quite different, were the same species.

4. Müller described some freshwater species in the *Zoologia*, like these anostracan fairy shrimps (*Chirocephalus diaphanous*) that live in ephemeral pools where their eggs withstand desiccation until the pools refill.

5. The orange-footed sea cucumber (*Cucumaria frondosa*) is an abundant species in the North Atlantic. Müller's dissection shows the curious highly branched, single gonad characteristic of holothurian echinoderms.

Zool. d. T. IV.

2.

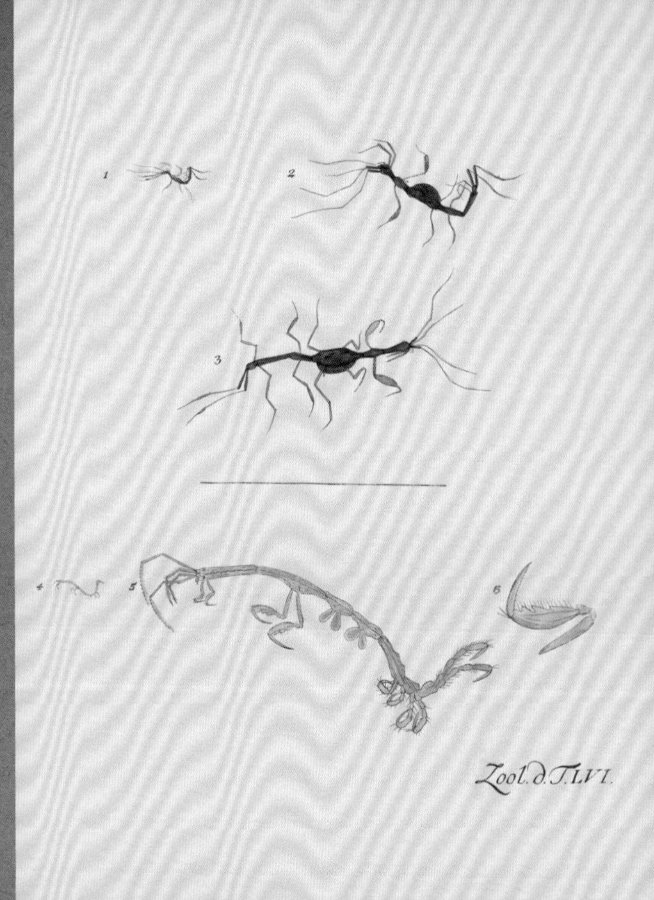
Zool. d. T. LVI.

3.

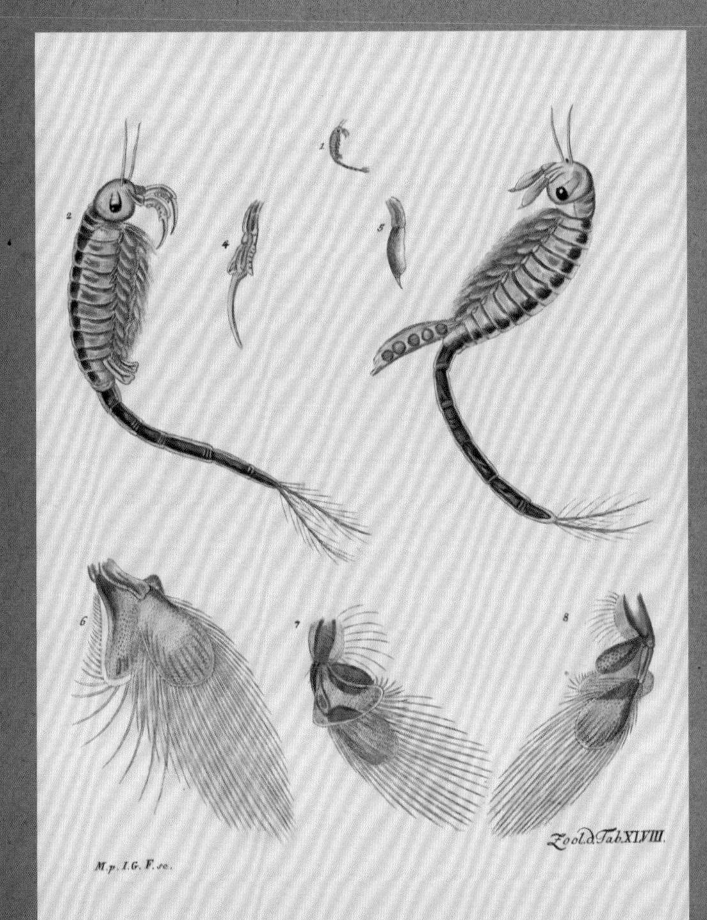
M. p. I. G. F. sc.
Zool. d. Tab. XLVIII.

4.

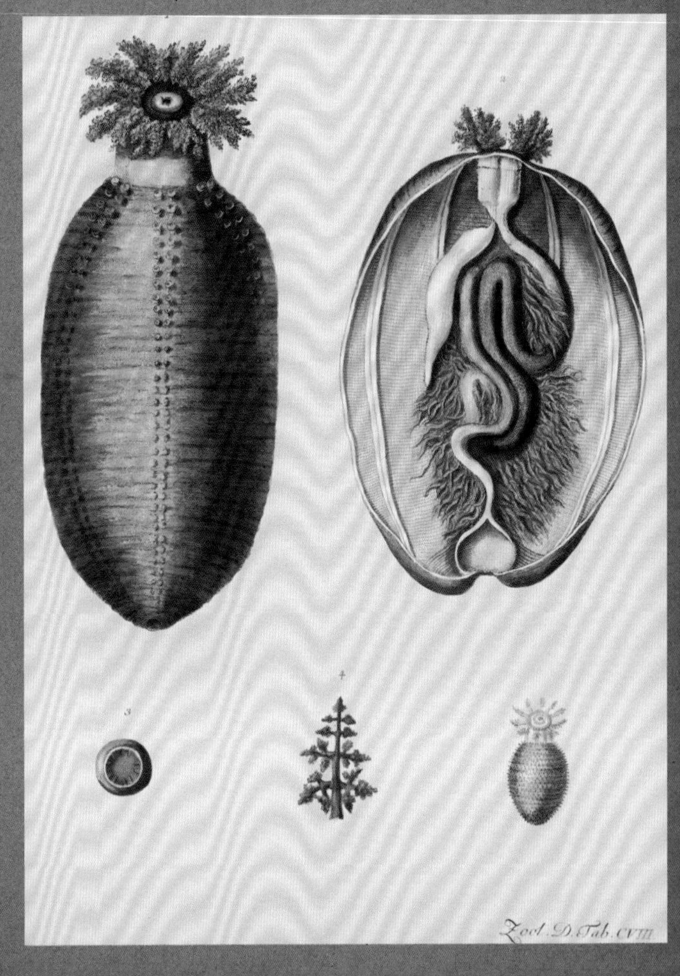
Zool. D. Tab. CVIII

5.

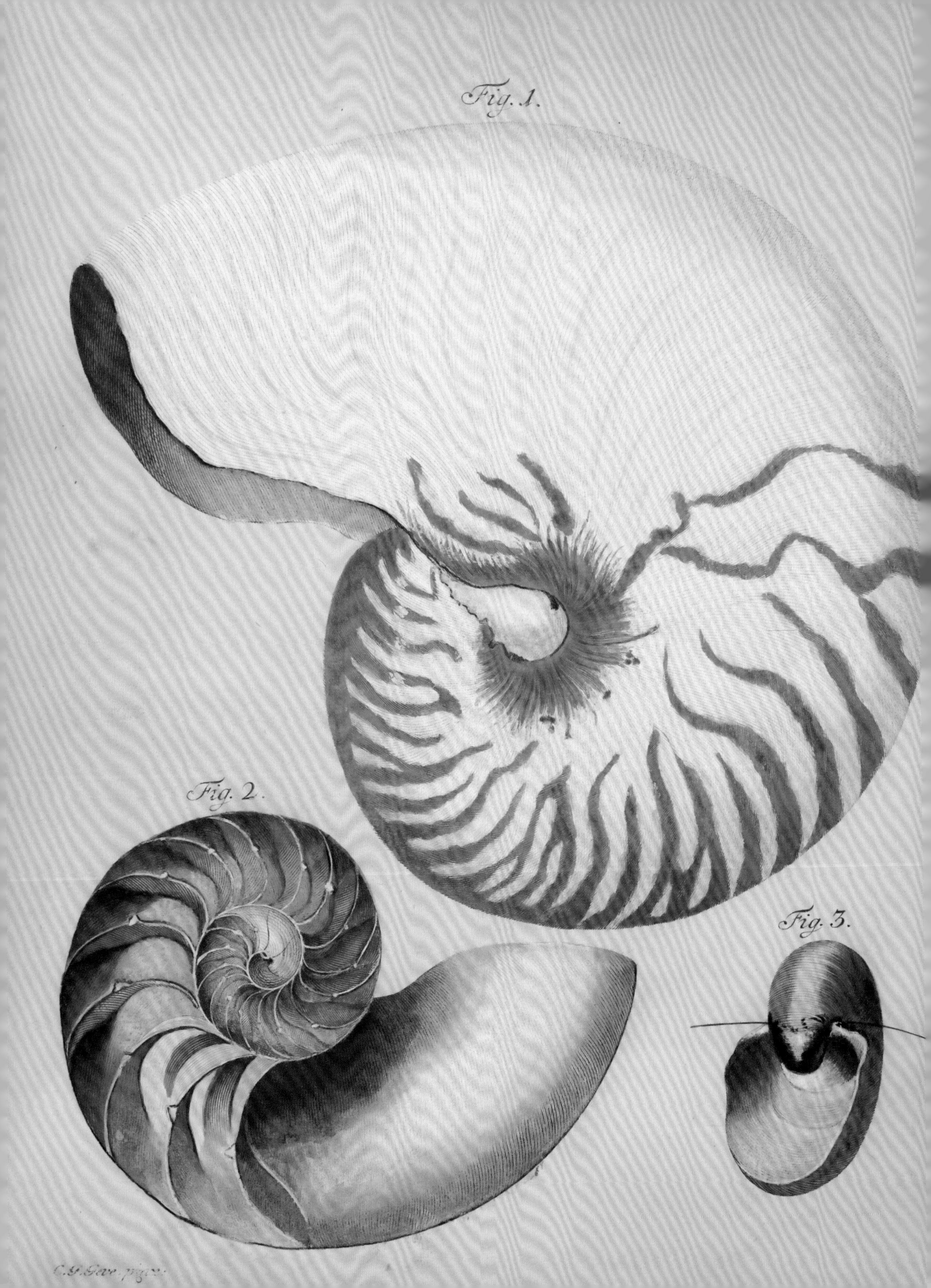
Fig. 1.
Fig. 2.
Fig. 3.
C.G. Geve pinx.

The Painter's Nautilus

Author
Nicolaus Georg Geve

Title
Belustigung im Reiche der Natur. Erster Band aus den Papieren des Verstorbenen vollendet durch Johannes Dominicus Schultze.

(Amusement in nature's kingdom. First volume of the papers from the late Nicolaus Georg Geve completed by John Dominicus Schultze.)

Imprint
Hamburg: Bey den Gebrüdern Herold, 1790

1. The beautiful color and symmetry of *Nautilus* shells has fascinated artists and mathematicians over the ages and made the shells much sought-after items for Renaissance collectors.

Very little is known about the eighteenth century Hamburg artist and painter Nicolaus Georg Geve, and neither the date of his birth nor his death has been recorded. We do know that he had traveled to Copenhagen to become a student of Jean Samuel de Wahl, painter to the court of Denmark and director of the Danish Museum of Curiosities in Copenhagen, and returned to Hamburg to work as an artist. It is likely that it was as an apprentice to de Wahl that Geve acquired his skills both in the painting and the preparation of fine copperplate engravings of natural history objects, particularly of shelled mollusks and cephalopods. We also know that *Belustigung im Reiche der Natur*, published in 1790, was, as its full title implies, a posthumous work. This beautifully illustrated compendium, the first illuminated volume on shells to be published in Germany, was based on descriptions and meticulously prepared plates made by Geve before his untimely death.

In 1755 Geve had announced his intention to publish descriptions and engravings that would be produced in monthly series, ultimately to be compiled into four volumes. The first volume was to include all univalve shells; the second, bivalves and multivalves; the third, urchins and starfish; and the fourth, corals and marine plants. Unfortunately Geve was unable to complete this task, and while subscribers did receive some plates, the promised volumes never materialized and the project languished. After his death, thirty-three of Geve's copperplates and some descriptions were purchased from his heirs by the publishers, Brothers Herold, and the Hamburg physician and natural historian Johann Dominik Schultze (1752–1790) was commissioned to complete the work for publication. In the preface to the *Belustigung im Reiche der Natur*, Schultze is highly dismissive of the scientific content of Geve's descriptions but full of praise for the accuracy of his drawings and the elegance and naturalism of the color plates. Schultze added descriptions that were missing from the purchased materials, rewrote certain sections, and reorganized Geve's original classification to conform more with the Linnean standard of the day, and ventured to "place my name in the title of the book in recognition of the part played" (translated from the German preface).

Eighteen beautifully colored plates accompany the present volume, which corresponds in part to Geve's original conception of the univalve shells—primarily marine gastropods (snails) and some particularly stunning illustrations of shelled cephalopods. Forty-one years later Geve's fifteen remaining univalve plates were finally published in 1831 in an extensively revised volume under the title *Nicolaus Georg Gevens Conchylien-Cabinet* by Freidrich Bachman.

Pearly nautiluses are highly unusual relatives of squid, octopuses, and their allies. While many fossil nautiloids are known, the group is represented today by

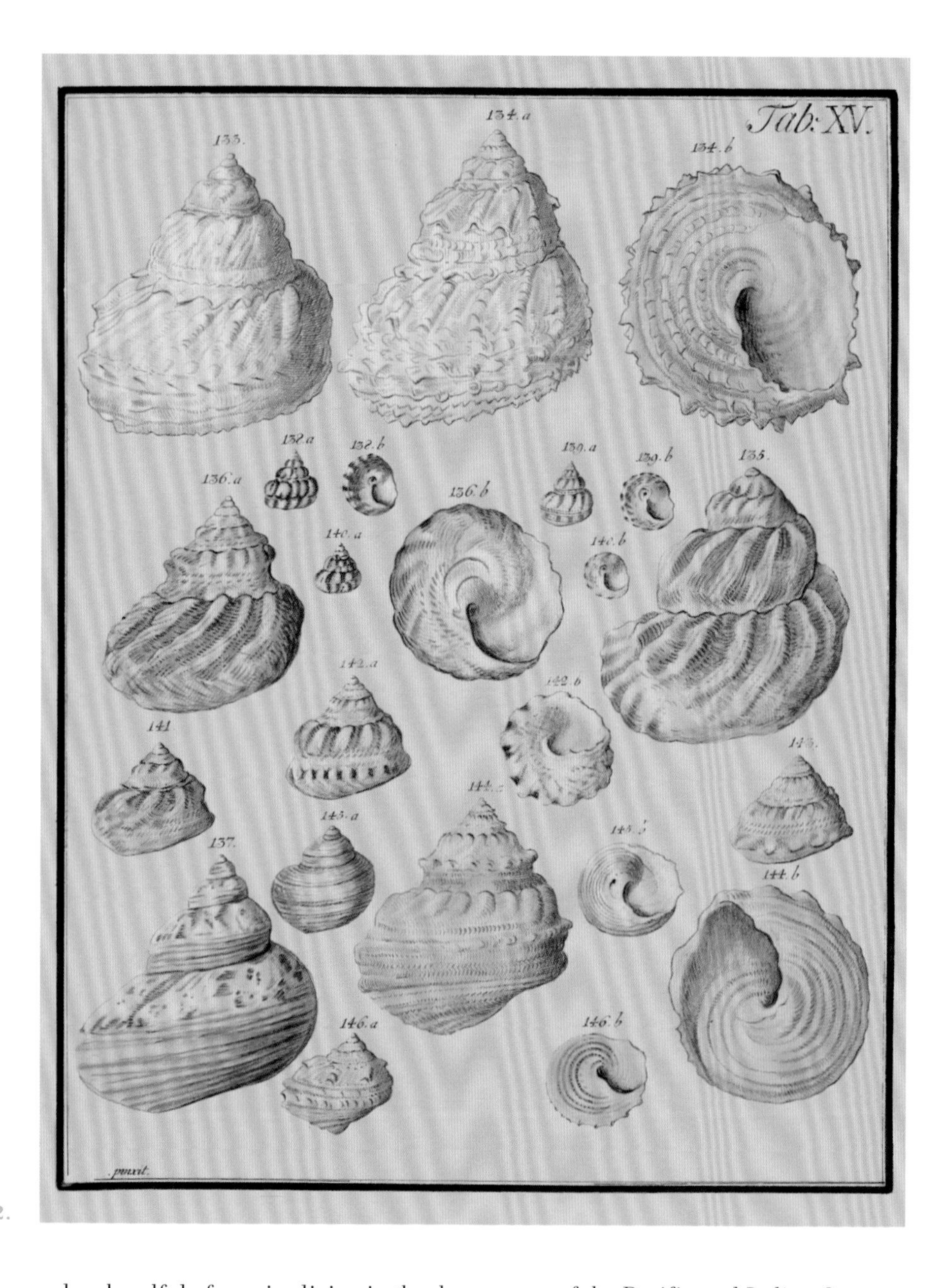

2.

only a handful of species living in the deep waters of the Pacific and Indian Oceans. The beautiful external shell, so characteristic of *Nautilus* species, is unlike that of any other living cephalopod and is internally divided into flotation chambers (phragmacone), which are linked by a tube (siphuncle) that allows gas in and out to determine buoyancy, with the animal living in the largest and most recently formed chamber. *Nautilus*, like most cephalopods, swims by jet propulsion and feeds using numerous tentacles (which lack suckers but are lined with adhesive ridges) to prey mainly on crustaceans. *Nautilus* shells have long been admired for their perfect symmetry (golden or logarithmic spirals) and color.

2. Most of the eighteen colored plates prepared by Geve depicted colorful gastropod seashells, such as this selection of the "buttoned moon shells."

3. The "shell" of the paper nautilus is actually an egg case of a pelagic octopus. The paper-thin egg case is secreted by the female argonaut octopus (*Argonauta argo*), and is wrapped around her body while she broods her eggs inside it.

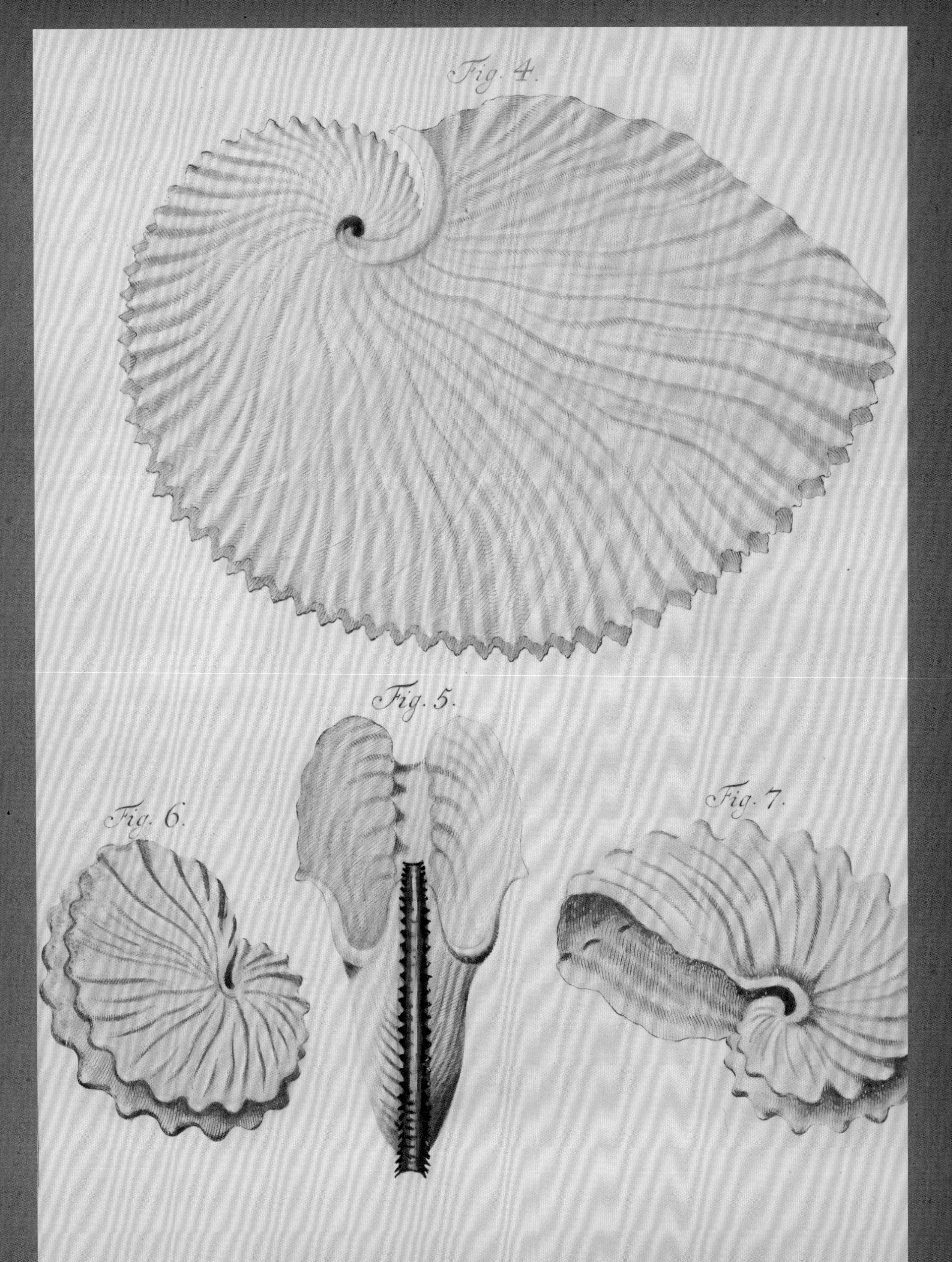
Fig. 4.
Fig. 5.
Fig. 6.
Fig. 7.

Testudo imbricata Linn.

The Turtles of North America and the Bahamas

Author
Johann David Schöpf (1752–1800)

Title
Historia testudinum iconibus illustrata

(Illustrated history of turtles)

Imprint
Erlangae: I. I. Palm, 1792–1801

1. Hawksbill sea turtles (*Eretmochelys imbricata*) have been hunted for their beautiful "tortoiseshell" for millennia. Today, they are protected under international conventions but, sadly, remain highly endangered worldwide.

Johann David Schöpf was one of the most accomplished of early travelers to venture through post-Revolutionary America. A physician by training and a naturalist by inclination, he lived at a time when it was still possible for one mind to grasp, and perhaps master, the essentials of all branches of science. He published influential papers on ethnology, meteorology, botany, zoology, and geology—all in addition to extensive medical writings.

Schöpf was born to an old and respected family in Wunsiedel, then part of the Bavarian principality of Ansbach-Bayreuth. He was educated by the family's private tutor and later attended the University of Erlangen. After matriculating, he studied briefly in Berlin, where, at the age of twenty-one, he was diagnosed with staphyloma and his right eye was removed. Writing to a friend, Schöpf blithely described the gruesome surgery and slyly commented that with his new "enameled eye" he could assure him "that I had much more success with the ladies than in former times." After travels in Switzerland, he defended his medical thesis in Erlangen, and then worked as a physician at a local orphanage. He was planning to travel to India when the margrave of Ansbach-Bayreuth summoned Schöpf to accompany his Hessian troops to America where they would fight alongside the British.

He arrived in New York in the summer of 1777 and was stationed on Staten Island where caring for the newly arrived German troops was grueling. He wrote that "this beautiful land was to become a cemetery to many of us in a short time." Dysentery and cholera took a dreadful toll during the humid New York summer and many troops died without ever encountering the enemy. Schöpf began researching his new surroundings and a series of essays, "On the Diseases in North America," "On the Climatic and Atmospheric Conditions of North America," and "The Effect of Poppy-Juice on Syphilis," were among the first of his works to appear in print.

In the spring of 1782, preliminary articles of peace were signed and the German troops returned home. Schöpf petitioned the margrave to remain in America and in July 1783 he traveled through Pennsylvania, Ohio, Maryland, Virginia, the Carolinas, Florida, and finally reached the Bahamas. Everywhere he engaged and quizzed the local populace, be they landowner, slave, or Native American. In addition to medical and ethnographic studies, Schöpf made meticulous observations and collections of biological and geological specimens, and in June 1784 he departed the Bahamas for Europe on board the ship *Hero*. The Atlantic crossing took thirty days and the *Hero* encountered many storms. Undeterred, Schöpf continued to make observations and was particularly struck by large numbers of Europe-bound sea turtles encountered on the homeward journey.

Back in Bayreuth, Schöpf was physician to the margrave and the royal family, but still found time to publish numerous works on his American discoveries. In 1787 an influential compendium on medicinal plants, the *Materia medica americana potissimum regni vegetabilis*, was published, followed a year later by a two-volume account of his journeys through North America (published 123 years later in English as *Travels in the Confederation*). By 1789 he was elected to the Imperial Academy, Leopoldina (formerly Academia Naturae Curiosorum), the same august body that had honored another great German naturalist, Georg Rumphius (see page 9). While Rumphius had been conferred the honorific "Plinius," Schöpf was to receive the epithet "Americanus II."

Schöpf continued his natural history compilations and, recalling observations made on the *Hero*, he began working up his notes on tortoises and turtles. He corresponded with numerous herpetological experts and in 1792 published the first part of the scholarly *Historia testudinum iconibus illustrata*. In the following years, additional parts were added, but sadly Schöpf did not live to see the publication of the fourth and final part, which appeared in print one year after his death in 1800. The *Historia* is a magnificent work in which Schöpf carefully described and discussed thirty-three kinds of tortoise and turtle, including three sea turtles that he found of particular interest. The thirty-four plates accompanying the work are by the famous illustrator and naturalist Friedrich Wilhelm Wunder (1742–1828), who worked directly from Schöpf's own drawings and from those Schöpf received from colleagues. In 1797 Schöpf was appointed president of the Medical College of Bayreuth, and in his last remaining years wrote mainly on medical topics, including a scathing commentary "On the effects of the medical system in the state, and its negligence in most German states," in which he decried the neglect of rural populations and the socially underprivileged. That socially progressive work was the last published in his lifetime and, sadly, this iconic figure of the German Enlightenment died of a throat malady at the age of only forty-eight.

Turtles (including tortoises and terrapins) are members of the Order Testudines (Chelonia) and about 260 species are known worldwide. Most live on land and a few in freshwater, but seven sea turtles live exclusively in marine habitats. Elegantly streamlined, with powerful flipperlike limbs that do not retract into their shells, sea turtles are supremely adapted to life roaming the oceans. Once hatched, male sea turtles never return to land, and females return only briefly to lay their eggs.

Their bony outer shells are formed from outgrowths of ribs and vertebrae and encase the limb girdles. The leatherback (*Dermochelys coriacea*) lacks a hard outer shell; instead, its bony plates are embedded beneath a thick, leathery skin, but other sea turtles have typically bony shells made up of dorsal (carapace) and ventral (plastron) plates. The carapace is covered by keratinous scutes that form the tortoiseshell. Tortoiseshell from the hawksbill turtle (*Eretmochelys imbricate*) was the most prized of all, and for centuries hawksbills were hunted primarily for their shells. Today trade in tortoiseshell is heavily regulated but, sadly, sea turtles are threatened by the destruction of their nesting beaches and a black market trade in their eggs and meat; but, most significant is their declining numbers due to accidental catch (bycatch) in long-line fisheries.

2. A hawksbill, viewed from below, shows the pattern of plates that forms the plastron. Highly modified for at sea, the turtle's large flippers cannot be retracted into the shell for protection.

3. Hatchling sea turtles, from top to bottom: hawksbill, green, and loggerhead. Females come ashore to nest, but once the eggs are deposited they return to sea leaving their eggs to incubate and hatch unguarded. The hatchling's sex is dependent on the temperature it incubated at.

4. Loggerhead sea turtles (*Caretta caretta*) get their common name for the large size of their heads and their powerful jaws. Here, Schöpf's illustration highlights the markedly reduced plastron of this species.

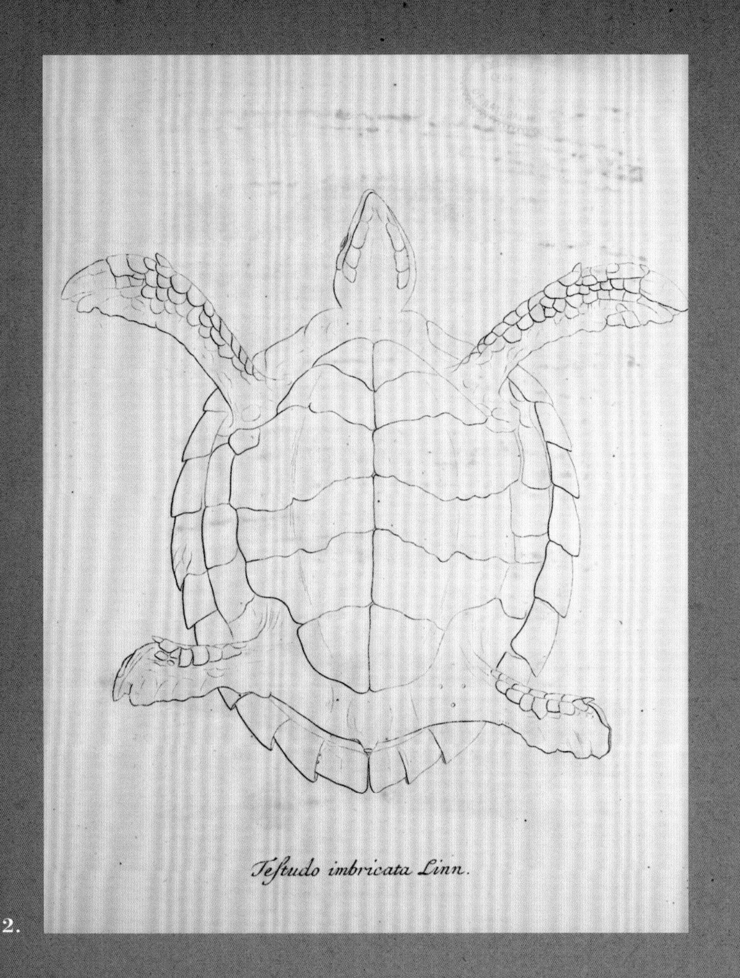
Testudo imbricata Linn.
2.

Tab. XVII.
Fig. 1. Testudo imbricata Linn. Fig. 2. Testudo Mydas Linn.
Fig. 3. Testudo Caretta Linn.
3.

Tab. XVI.
Testudo caretta Linn.
4.

Identifying the Unknown Fishes of India

Author
Patrick Russell (1727–1805)

Title
Descriptions and figures of two hundred fishes, collected at Vizagapatam on the coast of Coromandel

Imprint
London: Printed by W. Bulmer and Co., for G. and W. Nicol ..., 1803

1. This beautifully executed drawing of a white-spotted pufferfish (*Arthron hispidus*) named Bondaroo Kappa by Russell is one of many striking images that accompany this foundational work on Indian ichthyology.

Patrick Russell was a skilled physician who traveled widely and, in addition to undertaking important medical research, authored a series of seminal works on the flora and fauna of the Indian subcontinent. He was born and educated in Edinburgh, where his father was a prominent lawyer. Most of Russell's brothers and uncles were physicians, and he followed them into the medical profession. On graduating from the Royal Medical Society of Edinburgh, he left Scotland to join his older half brother, Alexander, in the Syrian Levant. Alexander was physician to the British Levant Company in Aleppo (Halab) and Russell planned to spend a few years there honing his medical skills. Three years later, in 1753, when Alexander returned to London, Patrick replaced him as company physician.

While in Aleppo, Russell made a particular study of the bubonic plague, a recurrent and devastating scourge, and years later he summarized his experiences in an influential medical volume, *A treatise of the plague*, published in 1791. Russell's medical attentions, rendered at great personal risk of infection, and his compassionate treatment of patients "equally conspicuous to the natives of all ranks, as to the gentlemen belonging to the English factory" gained him the affection of the local populace, whose languages he mastered and customs he respected.

Russell shared his brother's enthusiasm for the region's natural history, made numerous observations and collections, and sent regular notes back to Alexander, who was compiling a second edition of his *Natural history of Aleppo and parts adjacent*, a work that Russell was to completely revise and publish twenty-six years after Alexander's death. After nearly twenty years in Aleppo, Russell returned to Edinburgh, and then went to London, where he established a thriving medical practice. In 1777 he was recognized for his work in Aleppo by the Royal Society of London and was elected a fellow.

In 1781 his brother Claud was ailing and the family decided that Russell, now a fifty-five-year-old bachelor, should accompany him to India. Claud had been appointed chief administrator at the newly established East India Company station at Vizagapatam (Visakhapatnam) in coastal Madras Province and Russell, uprooted from his London life, sailed to India to care for his young brother. However, true to form, Russell arrived in Vizagapatam far from resentful and instead embraced the extraordinary opportunities afforded him by the relocation. He immediately began documenting the flora and fauna of the virtually unknown region, and when the company botanist John Gerard Koenig (1728–1785), with whom Russell had established a close friendship, died, Russell was asked by the governor of Madras to replace him. Russell thrived in the position and worked assiduously gathering specimens and local knowledge. In addition to botanical studies, he became increasingly interested in the region's snakes, particularly the venomous snakes that were so plentiful and which caused such suffering to the local inhabitants and company workers. By 1787 he had produced a practical guide for distinguishing the poisonous snakes of India, which was widely disseminated by the East India Company to its employees.

In addition to botanical and herpetological studies, Russell began pioneering work on coastal fishes. At the time, there was virtually nothing known of Indian fishes and Russell's London friend, the famed botanist Sir Joseph Banks (1743–1820), urged him to make an ichthyological study. This he did, and the result of "the fruit of many laborious hours, when disengaged from other pursuits of natural history" was to become his magnificent two-volume *Descriptions and figures of two hundred fishes, collected at Vizagapatam on the coast of Coromandel.* The accompanying 198 engraved plates, many of which were aquatint etchings, were drawn by a "native painter" who "learned in short time to delineate so accurately the parts pointed out to him, that his figures, howsoever deficient in art and grace, may in general be relied on in respect to fidelity in representation."

Russell had originally intended to have the illustrations colored from life but lamented that the hot climate caused the fishes' colors to fade and "escape while the painter is adjusting his palette" and decided instead to make detailed written color descriptions. Russell deposited his fish specimens at the company's museum at Madras, but retained the illustrations with the intention of completing the work at a later date.

After seven years in India, Russell returned home and set to work writing up his many discoveries, and in the ensuing years he published numerous groundbreaking studies. Although Russell is best known today for his pioneering treatises on venomous snakes, their habits and venoms, his magnificent *Descriptions and figures of two hundred fishes,* published in 1803, remains foundational in the history of Indian ichthyology.

At the age of seventy-eight, Russell died in London after a short illness. A practical clinician to the end, he instructed his executors to inter him in the nearest burial ground with minimal pomp, and most particularly not "within the walls of any place dedicated to public worship." After many years of treating infectious diseases, he considered the practice of burying bodies in churches "useless to the dead and prejudicial to the living."

Many of the drawings accompanying Russell's work are rather stylized, yet it is possible to easily identify most of them. The Bondaroo Kappa, for example, is quite recognizable as *Arothron hispidus,* the white-spotted pufferfish. It belongs to the Family Tetraodontidae, named for the four (tetra) toothlike (odont) structures that form the modified jaws of these fishes, which are used to crack open their prey, such as corals, crustaceans, and mollusks. Puffers have a novel defense mechanism and can rapidly fill their highly elastic stomachs with water (or air), blowing themselves up into a large balloon. This causes the many spines, which lie flat against the body when the fish is not inflated, to erect. The result is a large spiny ball hard for many predators to swallow. Many puffers have a second line of defense—tetrodotoxin, a potent neurotoxin concentrated in their livers and ovaries. While many large ocean predators, such as sharks, seem able to tolerate tetrodotoxin, it is highly toxic to humans. Interestingly, the first recorded case was by Captain James Cook (1728–1779), who reported classic tetrodotoxin poisoning among his crew after eating pufferfish caught on their voyage.

2. Although highly stylized, this image clearly depicts an electric ray. The dark edging to the pectoral fins suggest it is most likely the dark-finned numbfish (*Narcine maculata*) found in Indian coastal waters.

3. Russell's Mookarah Tenkee is a banded eagle ray (*Aetomylaeus nichofii*). This species has the very long tail characteristic of myliobatid eagle rays, but, unusually, lacks a sting.

4. In this plate, Russell's artist illustrates the undersurface of a pygmy devil ray (*Mobula eregoodootenkee*). This specimen is a male, and its claspers (tubular modifications of the pelvic fins) are used to transfer sperm during mating.

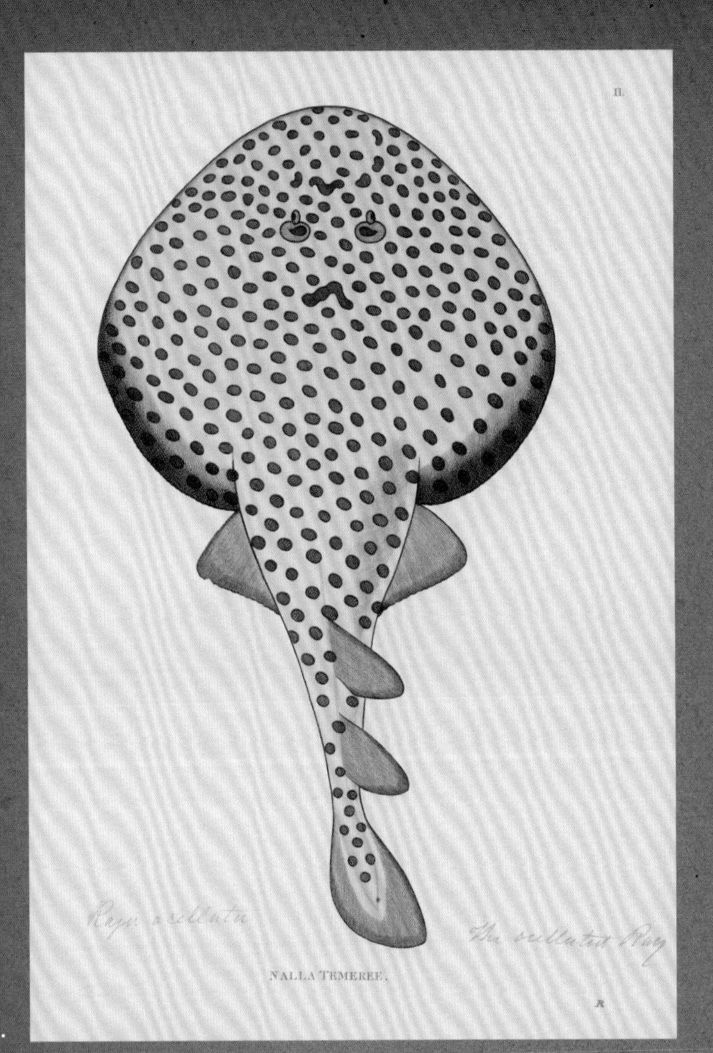
II.
Raja ocellata
The ocellated Ray
NALLA TEMEREE.

2.

VII.
MOOKARAH TENKEE.

3.

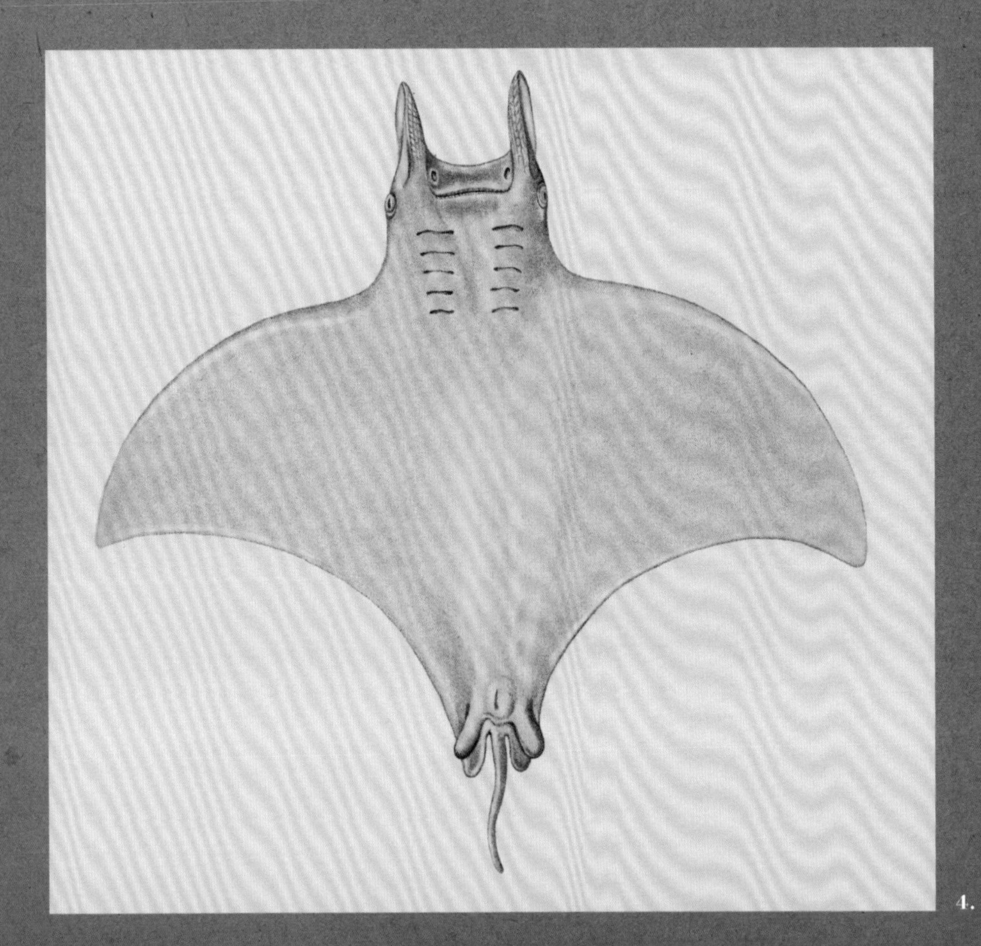

4.

2.
5.
4.

The Perilous Expedition to Australia's Coastline

Author
François Péron (1775–1810)

Title
Voyage de découvertes aux terres australes: exécuté par ordre du Gouvernement, sur les corvettes le Géographe, le Naturaliste, et la goëlette le Casuarina, pendant les années 1800, 1801, 1802, 1803 et 1804. Historique.

(Voyage of discovery to the southern lands: carried out by order of the Government on the frigates the Géographe, the Naturaliste, and the schooner Casuarina, during the years 1800, 1801, 1802, 1803 and 1804. Historical.)

Imprint
Paris: Imprimerie Impériale, 1807–1816

Figure 3
Author: Rene Primevere Lesson (1794–1849)
Title: *Voyage autour du monde, execute par ordre du Roi, sur la corvette de Sa Majeste, la Coquille . . . Zoology*
Imprint: Paris: A. Bertrand, 1838

1. Apart from the artistic merit of this plate, it is remarkable that Péron collected intact these highly fragile marine siphonophores, which easily break into pieces at the slightest touch.

François Auguste Péron was born in Cérilly, a small town in central France, and just thirty-five years later died of tuberculosis in that same town. Despite a tragically short duration, Péron's life was extraordinarily rich and eventful and he is remembered today as one of the pioneers of Australian natural history.

In the waning years of the French Revolution, Péron fought to defend France's occupation of Landau only to be wounded and imprisoned by Prussian forces. On returning to Cérilly in 1794, Péron, now invalided out of the army due to the loss of an eye, served for two years as town clerk before receiving a scholarship to study medicine in Paris. Three years into his medical studies, apparently due to failed romantic aspirations, he petitioned to join a cartographic survey of the coastline of Australia (New Holland) under the leadership of Nicolas Thomas Baudin (1754–1803).

Baudin's expedition sailed from Le Havre in October 1800 in two well-equipped corvettes, the aptly named *Géographe* and *Naturaliste*. Their mission was to chart the unknown lands of the Southern Ocean (Terra Australis) and to document the natural history of the seas and lands encountered. While primarily a scientific undertaking, it was launched at a time of intense competition between the French and the British for southern possessions. News of the French expedition stimulated the British to rapidly initiate a similarly charged mission led by the distinguished navigator and cartographer Matthew Flinders (1774–1814), who, following in the footsteps of Captain James Cook (1728–1779), had previously mapped part of the Queensland coast.

Péron had applied for the position of anthropologist on the French expedition but, somewhat unexpectedly, Baudin appointed him as apprentice naturalist, and Péron joined a team of more than twenty civilian scientists. Almost immediately his relationship with Baudin deteriorated and within a short time the two could express nothing but disdain for each other, an animosity that was to persist for both of their short lives. Baudin experienced similarly bad relations with others among the crew and on reaching Île de France (Mauritius), many scientists, officers, and seamen quit the mission.

After leaving Mauritius, shipboard conditions worsened as dysentery and scurvy began to take a heavy toll. By the time the ships approached Australia, just seven scientists and illustrators remained and Péron, now the sole naturalist in any condition to work, became the expedition's senior zoologist. Two other junior members, Nicholas-Martin Petit (1777–1805), a landscape artist, and Charles-Alexandre Lesueur (1778–1846), an illustrator whom Péron befriended, assumed central roles and worked with Péron to produce the exquisitely illustrated atlas accompanying the survey's reports.

Baudin had made a prescient appointment in Péron, who excelled in the tasks of botanical, zoological, meteorological, and anthropological observation, as well as in recording and collecting. Péron was also a pioneer of marine biology and took a series of important measurements of ocean surface and depth temperatures, which complemented work done during Cook's earlier Pacific explorations.

The soft-bodied zoophyte communities, so abundant in the Southern Ocean, particularly intrigued Péron and he produced some of the first observations on the biology and anatomy of these extraordinary animals. The plate accompanying this essay, artfully arranged and rendered by Lesueur, features a selection of hydrozoan siphonophores, illustrating the intricate beauty of these oceanic drifters.

Despite much travail, Baudin surveyed sections of the "unknown coast" of southern and western Australia, and very large collections of specimens were made. In April 1802 the *Géographe* and *Naturaliste* sighted the British ship, *Investigator*, captained by Matthew Flinders, who by this time had made excellent progress in his efforts to circumnavigate Australia—proving for the first time that it was a continent-size island. The meeting was amicable and Flinders informed Baudin of the existence of nearby Kangeroo Island, where he could find much-needed fresh provisions and water. Flinders designated the site of this meeting Encounter Bay, a name it bears to this day.

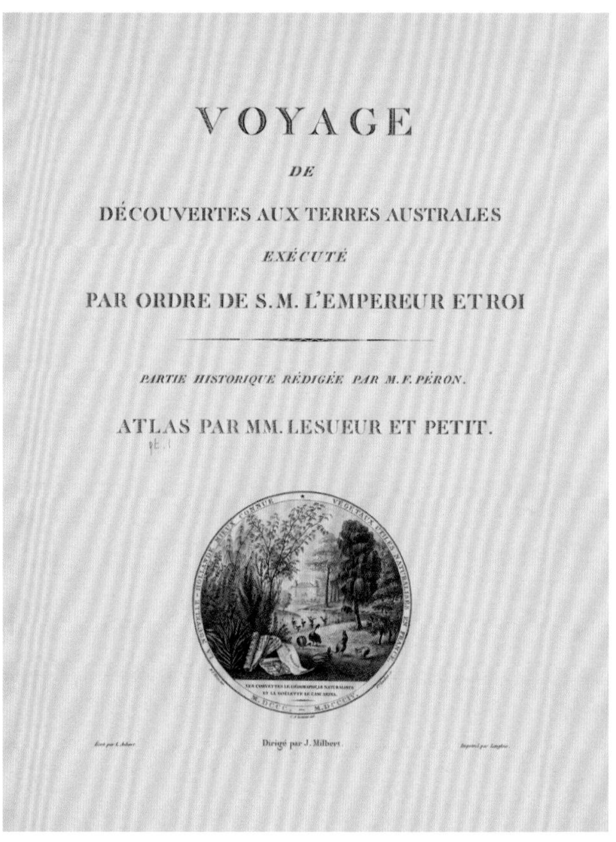

VOYAGE

DE

DÉCOUVERTES AUX TERRES AUSTRALES

EXÉCUTÉ

PAR ORDRE DE S.M. L'EMPEREUR ET ROI

PARTIE HISTORIQUE RÉDIGÉE PAR M.F. PÉRON.

ATLAS PAR MM. LESUEUR ET PETIT.

Dirigé par J. Milbert

2.

2. The title page of Lesueur and Petit's Atlas of illustrations.

3. This magnificent plate, prepared by the engraver Jean Loius Coutant, is based on a specimen collected during the Voyage of the *Coquille*. The species depicted is *Physalia physalis* and is distinguished from the Indo-Pacific species *Physalia utriculus* illustrated by Péron, which has a single long fishing tentacle.

On reaching Port Jackson (present-day Sydney), Baudin sent the *Naturaliste* back to France heavily laden with scientific cargo. The *Géographe* remained at port and a smaller, locally built schooner, the *Casuarina*, was purchased. Expedition cartographer Louis de Freycinet (1779–1841) was assigned the *Casuarina* for in-shore survey work, a task not suitable for the larger *Géographe*. Work continued, but by July 1803 many of the crew were ill, Baudin seriously so, and the survey was abandoned.

The ships returned to Mauritius, where Baudin succumbed to tuberculosis and the *Casuarina* was abandoned. Some months later, the *Géographe* sailed for France, arriving home on March 23, 1804. Shortly thereafter, Matthew Flinders, who had been forced to abandon the *Investigator* in Port Jackson and to attempt the return journey in the badly damaged schooner *Cumberland*, limped into port for repairs. Previously, and in thanks for the kindnesses received from Flinders and the British colonists at Port Jackson, Baudin had requested that Mauritius render service to any British ship forced to call at this French possession. But by the time of Flinders's arrival, Britain and France were once again at war and he was immediately arrested. Despite petitions for his release from both the English and French governments—even from Napoleon himself—Flinders was held on Mauritius for six years. He finally arrived in Britain in October 1810, some three years after Péron's first great volume was published. So it was that the French were able to publish the first detailed chart of Australia, although most of the cartography had been Flinders's work.

Physalie de L'atlantide. A.Ventouses grossies.
Physalia atlantica, Less. (Physalia pelagica, Lamk.)

Lesson et Bessa pinx. *De l'impr.^ie de Rémond.* *Coutant sculp.*

3.

Meanwhile, Péron and Lesueur, based at the Paris Natural History Museum, had set to work refining their field sketches and sorting through the more than one hundred thousand specimens and artifacts assembled during the voyage. In 1806 Emperor Napoleon I gave permission for the findings of the Baudin expedition to be published, and Péron and Lesueur were provided state pensions to support their efforts. In 1807 the first volume of Péron's *Voyage de découvertes aux terres australes: exécuté par ordre du Gouvernement, sur les corvettes le Géographe, le Naturaliste, et la goëlette le Casuarina, pendant les anneés 1800, 1801, 1802, 1803 et 1804,* accompanied by forty of Lesueur's stunning plates and many of Petit's superb landscape drawings, was published. But before completion of the second volume, ill health forced Péron back to Cérilly, where at the age of only thirty-five he succumbed to tuberculosis and died.

After Péron's death, Lesueur worked under the direction of their former shipmate, the cartographer Louis de Freycinet, and a second volume, principally concerned with cartographic and geographic findings, was published in 1816.

Péron had died before he was able to fully study the treasure trove of materials he had assembled, but his indefatigable scientific leadership and insightful observations are widely acknowledged today and he is rightfully recognized as one of the pioneers of Australian natural history.

The term "zoophytes" was applied by early workers to a broad mix of soft-bodied marine invertebrates, many of which are now grouped together with jellyfish (see page 140), box jellies (see page 49), anemones (see page 92), and corals (see pages 52 and 97) in the phylum Cnidaria. All of the zoophyte species illustrated in the accompanying plate are siphonophores, members of the cnidarian Class Hydrozoa and, although many superficially resemble jellyfish, they are actually colonies made up of many individual animals (zooids) specialized for different functions (feeding, defense, and reproduction). Each zooid is so intimately associated with all the others that the colony has the character of a single organism.

Most siphonophores swim suspended in the water column of the open ocean, and are long and gelatinous. Some can grow to 50 meters (c. 160 feet), making them among the longest animals on the planet. The infamous Portuguese man-of-war (*Physalia physalis*) is also a siphonophore (not a true jellyfish) that floats on the ocean surface. The "sail" of the man-of-war is actually an air-filled zooid (pneumatophore) and the long trailing tentacles (dactylozooids), armed with stinging, venom-filled nematocysts (stinging cells), are used to catch fish and plankton prey. Despite their stinging defenses, *Physalia* species are heavily preyed on by loggerhead sea turtles (see page 35), and are a particular favorite of the sea swallow (*Glaucus atlanticus*), a floating (pelagic) nudibranch sea slug (see page 147). That Péron chose to illustrate this voracious molluscan predator on the same plate as its favorite siphonophoran prey nicely illustrates his understanding of the ecological associations of many of the organisms he studied.

4. Lesueur's meticulously detailed sectional compositions of the coastlines of New Holland (western Australia) and Diemen (Tasmania) beautifully evoke the spectacular topology and grandeur of these newly discovered southern lands.

C. A. Lesueur del. *J. Milbert direx.* *Fortier sculp.*

TERRE DE DIÉMEN ET NOUVELLE-HOLLANDE.

1. *Mewstone.* (a.) *Îles de Witt.* (b.)
2. *Île Tasman.* (c.)
3. *La Piramide.* (d.) *Groupe de Kent.* (e.)
4. *Vue du Promontoire de Wilson.* (f.)
5. *Vue d'une partie de la côte Occidentale de l'Île Decrès : Cap Borda.* (g.) *ravine des Casoars.* (h.)

De l'Imprimerie de Langlois.

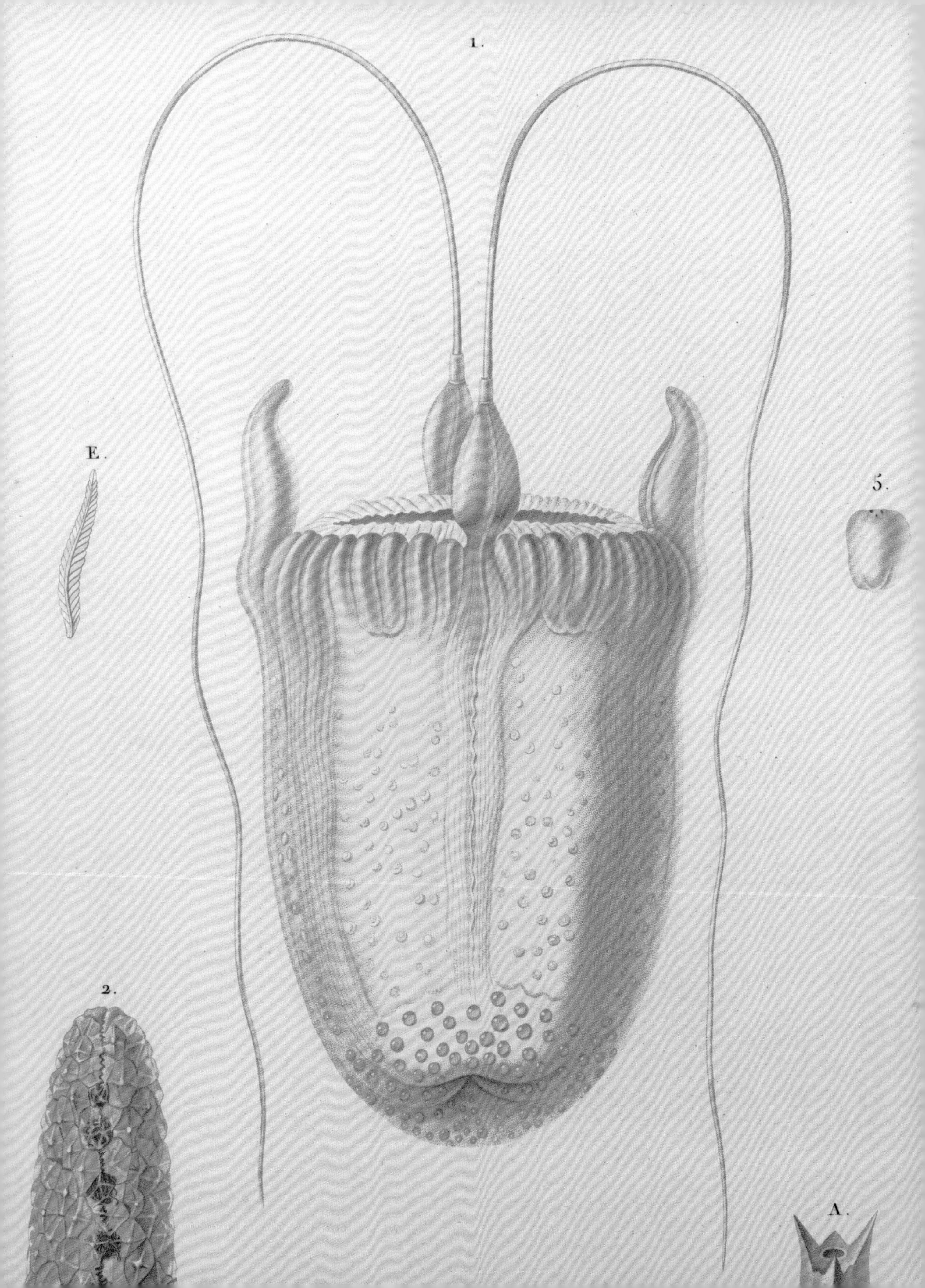
1.
E.
5.
2.
A.

Lesson Learned: Box and Comb Jellies

Author
René Primevère Lesson (1794–1849)

Title
Voyage autour du monde, exécuté par ordre du Roi, sur la corvette de Sa Majesté, la Coquille, pendant les années 1822, 1823, 1824, et 1825. Zoologie.

(Voyage around the world, carried out by order of the King, on His Majesty's frigate, the Coquille, during the years 1822, 1823, 1824, and 1825. Zoology.)

Imprint
Paris: A. Bertrand, 1838

Figure 3
Author: Louis Isidore Duperrey (1786–1865)
Title: *Voyage autour du monde, execute par ordre du Roi, sur la corvette de Sa Majeste, la Coquille . . . Histoire du voyage.*
Imprint: Paris: A. Bertrand, 1826

1. Among the most venomous of marine animals, cubozoans (box jellies and sea wasps) are agile swimmers, capable of actively preying on zooplankton and small fishes, which they immobilize with their potent venom.

The French defeat at Waterloo signaled the end of the long Napoleonic Wars, and once again Europe's attention turned from naval battles to naval exploration. For France, so militarily weakened after Napoleon's disastrous invasion of Russia, there was strong motivation to reinvigorate the nation's tradition of global exploration. And within short order, a stream of French "voyages of discovery" were launched with government sponsorship, to much public acclaim. One of these was the voyage of the *Coquille* under the command of Louis Isidore Duperrey (1786–1865), with the indomitable Jules Sébastien César Dumont d'Urville (1790–1842) as second in command. D'Urville, the more talented naval navigator, was destined to become one of his country's most esteemed explorers, regarded by many as France's own "Captain Cook." D'Urville was also an accomplished botanist who participated actively in the scientific activities of the *Coquille*, and on all subsequent expeditions. In later years, he eloquently wrote that "nothing is nobler or worthier of a lofty spirit than to devote one's life to the pursuit of knowledge. That is why my inclination urged me to voyages of discovery rather than to the purely fighting Navy."

Accompanying Duperrey and d'Urville on the *Coquille* were the young naval surgeon and pharmacist René Primevère Lesson and his fellow surgeon Prosper Garnot (1794–1838). Both men, although medically trained naval officers, had considerable knowledge of natural history, particularly that of mammals, birds, reptiles, and amphibians. However, in the preface to the first volume of the voyage's reports, Lesson alludes to the decision of the naval ministry not to include professional naturalists on board the *Coquille*, but rather to rely on the ship's surgeons. Just before departure, orders were sent that d'Urville was to take the botanical and entomological (insect) focus and, as Garnot had requested to work exclusively on mammals and birds, Lesson was left to fill in all the remaining taxonomic gaps himself.

The *Coquille* sailed from Toulon in 1822 on a voyage of some 117,500 kilometers (c. 73,000 miles)—rounding South America on passage to Australia and New Zealand, then across the South Pacific to the Cape of Good Hope, skirting the African coast before arriving back in Toulon in 1825. Out of sight of land for much of the voyage, Lesson was to recall that for long periods, "Phaetons [tropicbirds], flying fish, and shark often made a good appearance, as well as clumps of grasses carried along by the tides. These floating fields with their new forms of marine life would have been a great discovery for a naturalist. During these calm days, we also caught a good number of strange zoophytes: jellyfish, physalias, and an unusual ribboned beroid."

Lesson was aware that sea life would be an important focus and, although he appears to have been insecure in his abilities to adequately cover the field of invertebrate zoology, he nonetheless made numerous notes and sketches of many curious, soft-bodied zoophytes (see page 44) encountered, while making good collections of them. Although far from an expert on these animals, Lesson took sole responsibility for writing the final account of zoophytes in the second volume of the expedition's reports. He was also a key player in the compilation of the beautifully illustrated zoological atlas that accompanies the two volumes of *Voyage autour du monde, exécuté par ordre du Roi, sur la corvette de Sa Majesté, la Coquille*. Four large color plates in the atlas illustrate soft-bodied zoophytes, each apparently based on Lesson's original drawings.

Interestingly, unlike François Péron, who accompanied Baudin on his explorations of the southern ocean, Lesson was less familiar with zoophytes and their classification. As a result, a number of specimens were incorrectly identified or assigned to groups that are no longer recognized. Nonetheless, Lesson's notes and illustrations are of considerable interest, and some are quite beautifully rendered. One of these, shown in the accompanying plate, he named "Bourse de Vénus" (Venus's purse). This is clearly a box jelly (Cubozoa), but illustrated on the same plate are sponge larvae (Porifera; see page 100), a colonial siphonophore (see page 44), and a jellyfish (see page 140). A similar taxonomic mélange is represented on another plate, where the central image is of a comb jelly (Ctenophora), with another ctenophore drawn above it, but also illustrated are a colonial siphonophore and a string of mollusk eggs. On yet other plates, Lesson mistakes jellyfish for ctenophores and ctenophores for mollusk eggs. With hindsight it is easy to be critical but, in truth, Lesson was well aware of the shortcomings of his knowledge of these animals and under the circumstances he performed a remarkable task. His more authoritative zoological contributions are found in the accounts of the terrestrial vertebrates he authored.

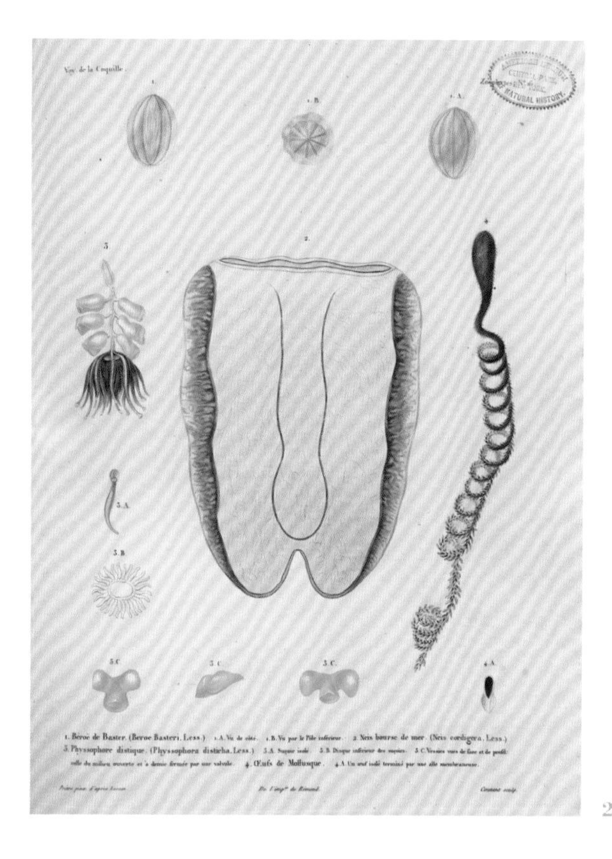

2. In the center of this plate is a ctenophore (*Neis cordigera*) that Lesson had collected while the *Coquille* was docked at Port Jackson (Sydney). Lesson's *Neis* is one of the largest beroids, occasionally attaining lengths of almost 30 centimeters (c. 1 foot).

Early in the voyage, Garnot contracted dysentery and by the time the *Coquille* reached Port Jackson (Sydney), his condition had so worsened that he was forced to leave the expedition. He left on an English vessel, the *Castle-Forbes*, taking with him many crates of specimens. The *Castle-Forbes* was wrecked off the Cape of Good Hope in July 1824; Garnot survived, but the specimens were lost. Lesson later wrote, "in just one day the perseverance and care of a full year [of collecting] were lost." Nonetheless, one year later, the *Coquille* returned to France with its hold filled with zoological specimens and artifacts. Lesson spent the following seven years preparing accounts of the vertebrates for the first volume of *Voyage autour du monde, exécuté par ordre du Roi, sur la corvette de Sa Majesté, la Coquille*, including many of Garnot's observations, and the volume was jointly authored. Lesson then completed the second volume, with assistance from a number of crustacean and insect experts, but authored the section on the zoophytes himself.

In the following years, Lesson continued to publish natural history works, but his naval career took precedence and he wrote a number of influential books on naval medicine. By 1835 he had become France's top-ranking naval pharmacist and was based at the naval port of Rochefort, the town where he was born and where he remained until his death in 1849.

3. Many beautiful renderings of places and peoples were included in the voyage reports, such as this delightful landscape of the main island of Bora Bora as seen from aboard the *Coquille* as it entered the lagoon.

Box jellies, often called sea wasps, belong to the Order Cubozoa and are close relatives of anemones and corals, hydrozoans, and jellyfish; together they form the large Phylum Cnidaria. Box jellies are easily distinguished from true (scyphozoan) jellies by their cube-shaped bodies. Under their bell is a distinctive flap (velarium), which allows them to concentrate and direct the flow of water as they swim, making them much faster than jellyfish. About thirty-six box jelly species are known, and they are found in most tropical and subtropical surface waters.

All cnidarians have tentacles covered with batteries of venom-filled nematocysts (stinging cells). However, box jellies have a different kind of venom that is particularly potent. They possess more numerous nematocysts and each tentacle can be armed with half a million stinging cells. While not all species pose a threat to humans, some certainly do, and the Australia sea wasp (*Chironex fleckeri*) is considered one of the most lethal animals in the sea. *Chironex* grow to about 15 centimeters (c. 6 inches) in length, with stinging tentacles extending nearly 3 meters (c. 10 feet). It is estimated that large individuals contain enough venom to kill sixty adult humans. Antivenom to the box jelly's sting does exist but needs to be rapidly administered as death can occur within seconds to minutes. But death from box jelly envenomation is relatively rare and most encounters result in only varying degrees of discomfort.

Another group of zoophytes illustrated by Lesson are comb jellies and, although jellyfishlike in appearance, these belong to a separate phylum, the Ctenophora. About 150 ctenophore species have been described; they live in all oceans, from the surface to the deep sea. Ctenophores are very distinctive in having "combs" of cilia that run the length of their bodies. These are iridescent, glowing red, green, or blue as the cilia beat and move the ctenophore through the water. Many ctenophores are also bioluminescent (see page 140) and, in some coastal regions, during the summer months they can become very abundant, lighting up the bays at night.

VUE DE L'ILE BORABORA.
(ILES DE LA SOCIETE)

3.

1

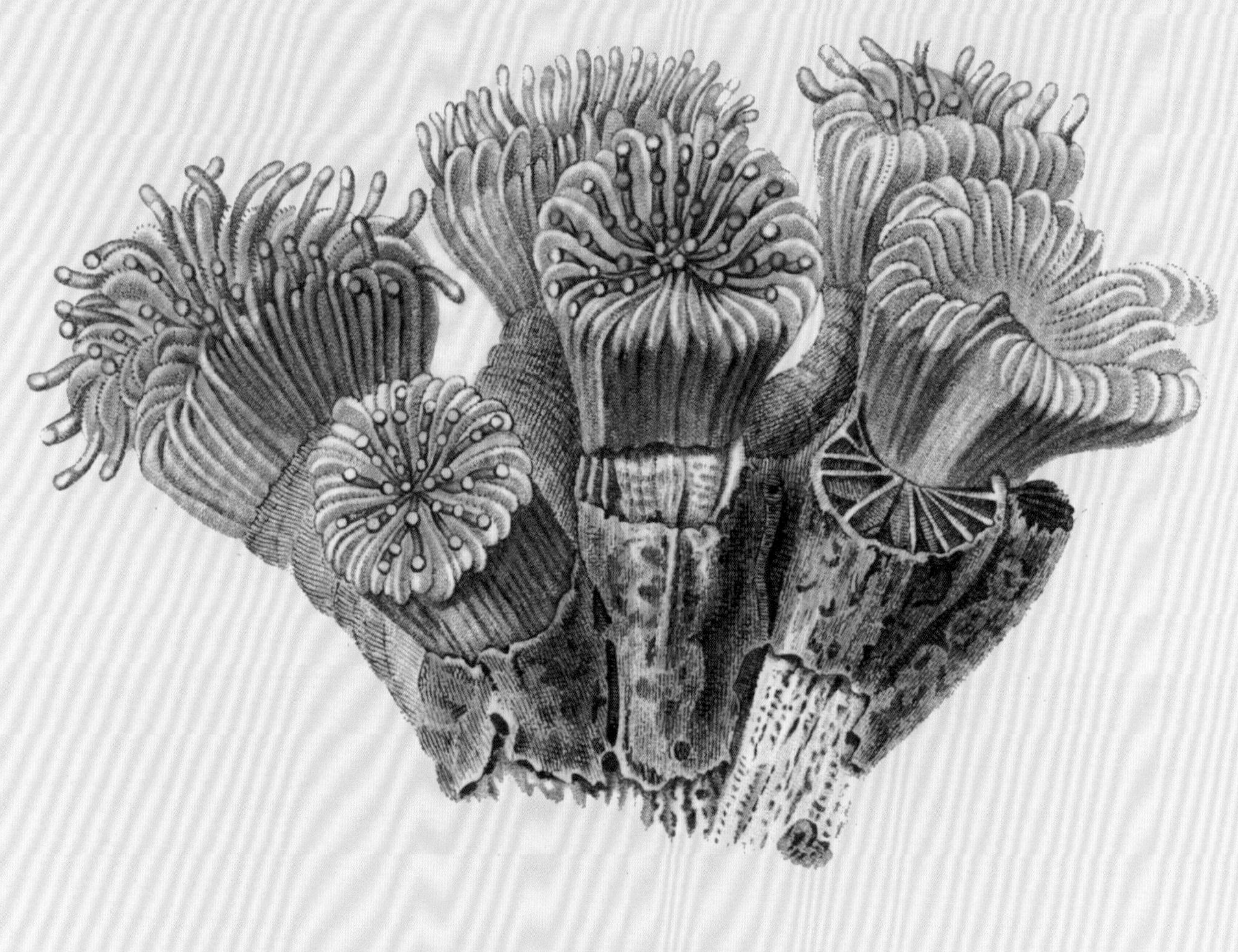

2

3

Quoy & Gaimard's Oceania

Authors
Jean René Constant Quoy (1790–1869) and
Paul Gaimard (1796–1858)

Title
Voyage de la corvette l'Astrolabe: exécuté par ordre du Roi, pendant les années 1826–1827–1828–1829, sous le commandement de m. J. Dumont d'Urville, capitaine de vaisseau. Zoologie.

(Voyage of the frigate Astrolabe: carried out by order of the King, during the years 1826–1827–1828–1829, under the command of Captain J. Dumont d'Urville. Zoology.)

Imprint
Paris: J. Tastu, 1830–1835.

1. The beautiful colors of many stony corals come not from the coral animals themselves but from photosynthetic algae (zooxanthellae) housed in their tissues. This relationship is mutually beneficial, as the coral provides protection while the algae supply oxygen and food for the coral polyps.

The brilliant French navigator Jules Sébastien César Dumont d'Urville (1790–1842) had served as botanist and second in command to Louis Isidore Duperrey (1786–1865) on the around-the-world voyage of the *Coquille* (see page 47). Despite Duperrey's admiration for d'Urville's botanical studies, the two men had become estranged and the younger naval officer was plainly anxious to establish his place in this golden age of French naval exploration. Ambitious and driven, just two months after the return of the *Coquille*, d'Urville petitioned the French Naval Ministry for a second voyage in the South Seas, but this time under his command. Recognizing d'Urville's abilities, the Crown, keen to maintain dominance over the British as both nations vied for naval supremacy, acceded. And in short order d'Urville was given command of the *Coquille*, renamed the *Astrolabe*, for an epic voyage that was to make his name as one of his nation's finest navigators and esteemed explorers.

D'Urville's mission was to chart unknown regions of Oceania, collecting samples and specimens en route, but also to determine the fate of the French aristocrat Jean-Francois de Galaup, comte de la Perouse (1741–1788), whose ships the *Boussole* and the *Astrolabe* (for which the *Coquille* had been renamed) were last sighted in March 1788 leaving Botany Bay, never to be seen again.

While d'Urville was confident he could chart unknown waters and determine la Perouse's fate, he understood the importance to the expedition's success of recruiting competent scientists, and it is no coincidence that three well-trained naturalists accompanied the *Astrolabe*. On board were the surgeon botanist Pierre Adolphe Lesson (1805–1888), the talented younger brother of René Primevère Lesson (1794–1849) who had sailed with d'Urville on the *Coquille*, and two zoologically trained naval surgeons, Jean René Constant Quoy and Joseph Paul Gaimard. Both were experienced naturalists who had worked together on the voyage of the *Uranie*, under the command of Louis de Freycinet (1779–1841) (see page 42). Originally hired as Quoy's assistant, Gaimard rapidly established himself as his equal and together they made many zoological discoveries, including the first tentative speculations as to how coral reefs and atolls could have formed in the middle of deep tropical oceans.

Gaimard toured the great museums of Europe, familiarizing himself with all that was known of Oceania, and when they set sail in April 1826, d'Urville's naturalists were prepared. Their voyage was to take nearly three years, charting much of New Zealand's poorly known coastline before exploring parts of New Guinea and many other islands in the vast ocean expanse that d'Urville named Melanesia and recognized as culturally distinct from Micronesia and Malaysia (regions he also named). And it was while exploring in the Melanesian Solomon Archipelago, among the reefs of Vanikoro, that d'Urville found the wreck of la Perouse's *Astrolabe*.

The local people told of two large ships that wrecked at Vanikoro and of some surviving crew who had built a boat and sailed away, but d'Urville was never able to find any trace of survivors. He erected a monument to la Perouse's expedition on Vanikoro before heading into Micronesia, where he mapped part of the Carolina Island chain and the Moluccas. When the *Astrolabe* sailed into Marseille on March 25, 1829, its hold was full of biological, geological, and ethnographic specimens, a haul so magnificent that the eminent Baron Cuvier (1769–1832) sought out Quoy and Gaimard to congratulate them on their remarkable work.

Between 1830 and 1834, d'Urville oversaw the publication of a magnificent fourteen-volume account of the *Voyage de la corvette l'Astrolabe*. The monumental work was accompanied by six beautifully illustrated atlases for which many of the nation's most accomplished natural history artists provided illustrations of specimens, landscapes, peoples, maps, and charts. Quoy and Gaimard reported on their zoological findings in four volumes and two large-format atlases. Many of the species illustrated, such as the stony corals that accompany this essay, were by the famed illustrator Jean-Gabriel Prêtre (1800–1840) and were often based on sketches made from life by Quoy, who was also a talented artist.

This pair of pioneering explorer/naturalists, whose names are forever linked in the taxonomic literature simply as "Quoy & Gaimard," discovered many new species in Oceania: a diverse array of groups including pinnipeds and cetaceans, fishes and mollusks, cnidarians and echinoderms, among many, many others. The daring and successful voyage of the *Astrolabe* was cause for national pride; d'Urville had successfully navigated and mapped vast swathes of Oceania, found the site of la Perouse's wrecked expedition, and safely returned with a tremendous natural history bounty and an ethnographic understanding of Oceania that would serve as the foundation for all expeditions that were to follow.

There are over three thousand species of stony coral (scleractinians) living mainly in shallow warm ocean waters. Quoy and Gaimard showed that these are the primary builders of the reefs and atolls found in such abundance in the warm South Seas. Unlike most soft corals (see page 22), stony coral skeletons form a hard aragonite (calcium carbonate) encasement around the living polyps. Although the polyps are very small, over time their colonies can grow into immense reefs that provide complex habitats for innumerable other marine species.

Despite sharing the name "coral," stony corals are actually more closely related to sea anemones than they are to soft corals. Sea anemones and stony corals are hexacoralians. Each coral polyp (or sea anemone) is ringed with six (or multiples of six) tentacles, while sea fans and soft corals (see pages 22 and 96) are octocoralians and their polyps are larger and usually bear eight tentacles.

2. Cuttlefish (Order Sepiida) are named for their buoyant internal cuttlebone, and Quoy and Gaimard illustrate the cuttlebones (left and right center) of these two attractive cuttlefish (*Sepia vermiculata*, above, and *Sepia papillata*, below) from the Cape of Good Hope.

3. *Sepioteuthis australis* (above) and *Sepioteuthis lessoniana* (below) are reef squids, but their large, rounded fins, extending around their bodies, give them a rather cuttlefishlike appearance; however, they are true squids (Order Teuthida), and as such, lack an internal cuttlebone.

4. This charming rendering depicts a large, male Australian sea lion (*Neophoca cinerea*) collected by Quoy and Gaimard along the coast of southwestern Australia.

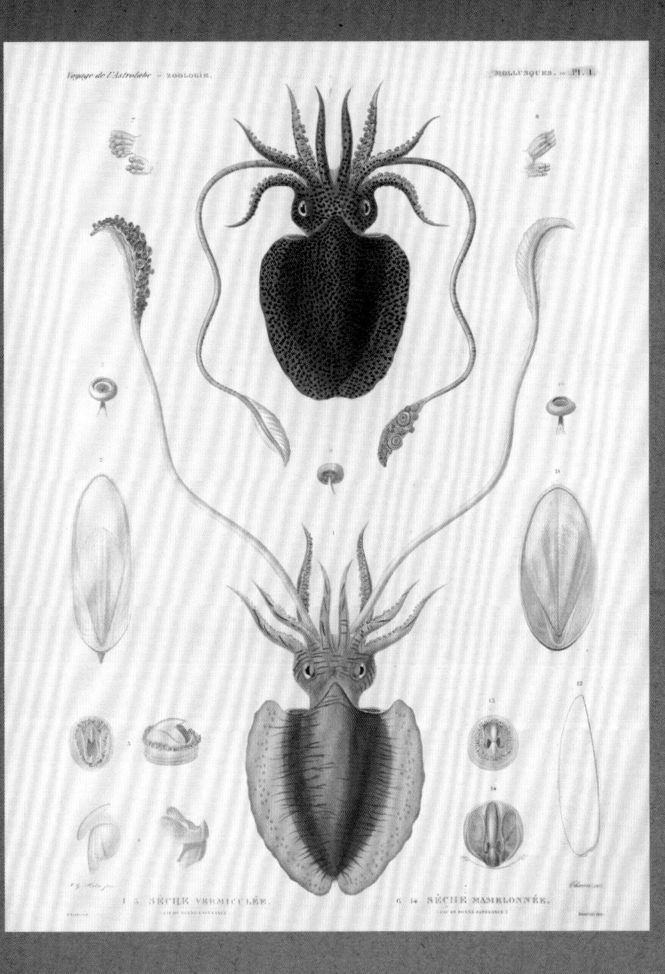
Voyage de l'Astrolabe – ZOOLOGIE.
MOLLUSQUES. – Pl. 1.
1 5 SÈCHE VERMICULÉE.
6 14 SÈCHE MAMELONNÉE.

2.

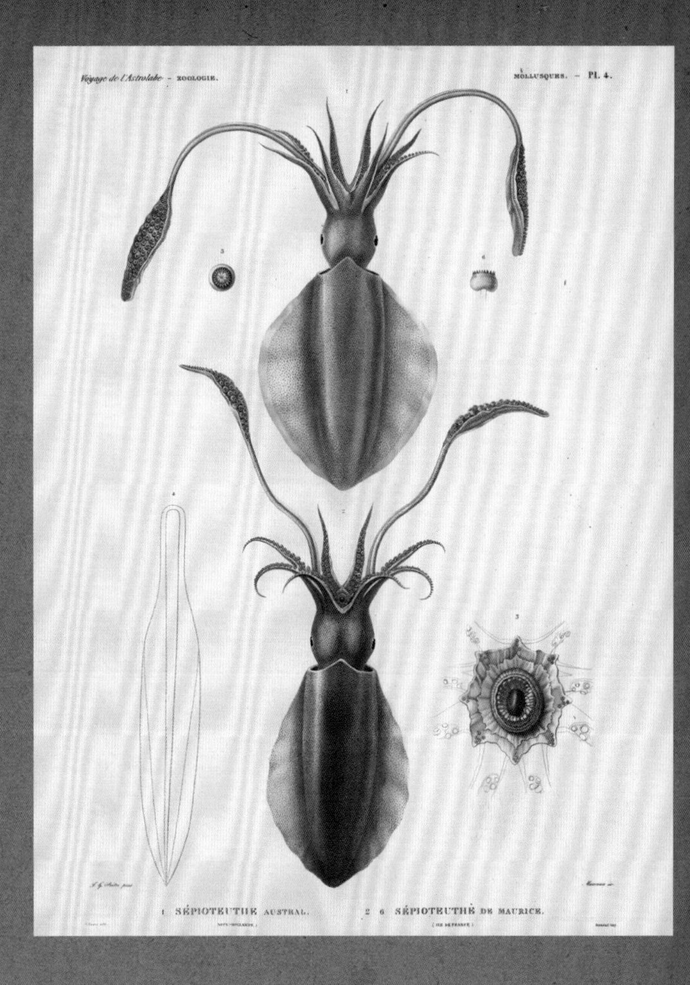
Voyage de l'Astrolabe – ZOOLOGIE.
MOLLUSQUES. – Pl. 4.
1 SÉPIOTEUTHE AUSTRAL.
2 6 SÉPIOTEUTHE DE MAURICE.

3.

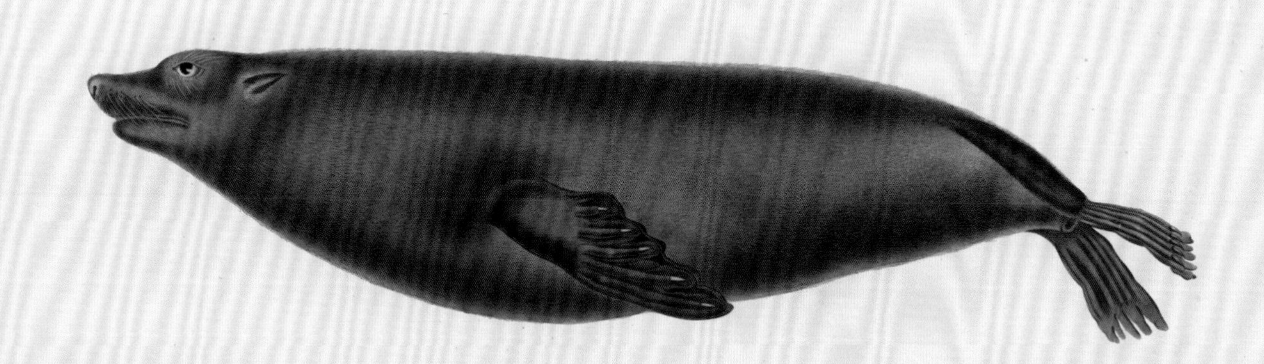
OTARIE CENDRÉE, MÂLE.
(NOUVELLE-HOLLANDE)

4.

Popularizing Natural History for Everyone

Author
William Jardine (1800–1874)

Title
The naturalist's library

Imprint
Edinburgh: W. H. Lizars, 1846

1. This charming Arctic scene depicts a family of bearded seal (*Erignathus barbatus*), an important food of polar bears. Pups are fast growing, adding an average 3.3 kilograms (c. 7 pounds) a day, and are ready to fend for themselves around twenty days after birth.

The naturalist's library was one of the most influential and widely disseminated natural history compendia published in the English language. Produced at a time of steam-driven technological innovation in printing and papermaking, the beautifully illustrated *Library* was produced at a fraction of the cost of previous works. With each of its volumes selling for just six shillings, by both price and design the *Library* was destined to reach a wide audience whose interest in natural history was burgeoning during Britain's rapid industrial growth and urbanization.

It was the entrepreneurial Edinburgh publisher William Home Lizars (1788–1859) who conceived of the *Library* with a vision to reach a large audience curious to learn of the many newly discovered species that were flooding into private collections, museums, and popular zoological gardens around the country, but with "shillings rather than guineas to spend." Lizars underwrote the ambitious project, but was confident of its success not the least because of the choice of his brother-in-law, the Scottish naturalist and aristocrat, Sir William Jardine, as series editor. Jardine, the seventh baronet of Applegirth in Dumfriesshire and hereditary laird of Spedlins Castle, was the epitome of an aristocratic British gentleman whose love of nature was kindled, as for so many of his time and class, by "rod and gun, and riding to hounds." But for Jardine, natural history, particularly the study of birds (ornithology), was a lifelong passion to which he made numerous scholarly contributions.

After medical studies at the University of Edinburgh Jardine traveled to Paris to continue anatomical training, but his father's death in 1821 demanded his return to Scotland. Despite many duties as heir to the baronetcy, Jardine published in 1825, with his friend Prideaux John Selby (1788–1867), the first of four volumes of the highly acclaimed *Illustrations of ornithology* and in 1831 helped found the influential Berwickshire Naturalist's Club, "instituted for the purposes of examining the Natural History and Antiquities of the county and its adjacent districts." The proliferation of naturalist's societies modeled on the Berwickshire Club is credited with forging the popular movement for the study of British natural history undertaken with such vigor throughout the Victorian era.

It was also at this time that Jardine recognized the importance of popularizing natural history beyond naturalist's clubs to reach those unable to participate directly and those disenfranchised by the high price of most illustrated natural histories. Jardine agreed to act as editor of *The naturalist's library* and, through his relationships with specialists and knowledgeable amateurs, was able to help Lizars find contributors to write individual volumes in their fields of animal study.

Between 1833 and 1843, forty volumes in the popular series were produced, and each was illustrated with at least thirty beautifully hand-colored plates engraved

2. Walrus (*Odobenus rosmarus*) have prominent tusks (modified canine teeth) surrounded by a broad mat of stiff, sensitive bristles that are used to locate food. They are deep divers and can remain submerged for as long as thirty minutes foraging for clams, which they can literally suck out of their shells.

by Lizars using a new steel engraving process that he had perfected. Lizars, considered one of the finest naturalist engravers of the early nineteenth century, was adamant that the illustrations should be of the highest quality so that "anyone unacquainted with zoology can obtain pleasure and profit from glancing at illustrations alone." He enlisted the services of accomplished natural history artists, including William John Swainson (1789–1855), Edward Lear (1812–1888), and William Dickes (1815–1892), but the man who contributed by far the most—a total of 545 signed plates—was James Hope Stewart (1789–?), who lived and farmed just six miles from the Jardine family seat at Jardine Hall. The stunning drawings set against beautifully rendered background landscapes that accompany this essay are the work of this poorly known but wonderfully talented Scottish amateur.

The *Library* was organized into four categories: Ornithology (the study of birds, in fourteen volumes), Mammalogy (the study of mammals, in thirteen volumes), Ichthyology (the study of fish, in six volumes), and Entomology (the study of insects, in seven volumes). Jardine himself wrote fifteen of the volumes and edited the remaining twenty-five, prefacing each with an essay on the life of a famous naturalist. For example, in volume 12, "Amphibious Carnivora; including the walrus and seals, and the herbivorous cetacea," Jardine provides an essay on François Auguste Péron (see page 41), while in volume 24, "Fishes, particularly their structure and economical uses," he wrote on Hippolito Salviani (see page 5).

Even while busy with *The naturalist's library*, Jardine continued to publish major ornithological works, but also turned his attentions to the study of British fishes, particularly members of the trout Family Salmonidae, and his beautifully illustrated *British Salmonidae* was issued in twelve parts between 1839 and 1841. In 1860, among many other duties, Jardine accepted appointment as Royal Commissioner of Salmon Fisheries, charged with investigating the causes of a precipitous decline in the British fishery.

Throughout a long and illustrious career, Jardine received honors from around the globe, and continued to play a central role in numerous scientific societies right up until his death at the age of seventy-four from a massive stroke while busy correcting proofs at his desk. Genial and approachable, and seemingly oblivious of social position, Jardine's life's work had the sole purpose of making "available to as many people as possible knowledge of the wonderful natural world about them" and, among his many achievements, there is little doubt that his labors on *The naturalist's library* did a great deal to attain that generous goal.

Seals, sea lions, and walruses belong to the mammalian Order Carnivora and are grouped together in the Pinnipedia, named in reference to their feather- or wing-like (*pinna*) flippered feet (*ped*). About fifty species of pinnipeds are placed in three separate families: the Phocidae or "earless" true seals, the Otariidae or "eared" fur seals and sea lions, and the closely related Odobenidae, which includes just the walruses. They range in size from about 1 meter (c. 3 feet) in length in the smallest Baikal seals, to true ocean giants like the southern elephant seals with some males almost 6 meters (c. 20 feet) long and weighing up to 4,000 kilograms (c. 8,800 pounds), making them the largest of living carnivores.

Although found in all the world's oceans, most pinnipeds prefer colder waters, where they spend most of their lives, but all must come ashore (or haul up on ice) to breed, birth, and nurse their young.

3.

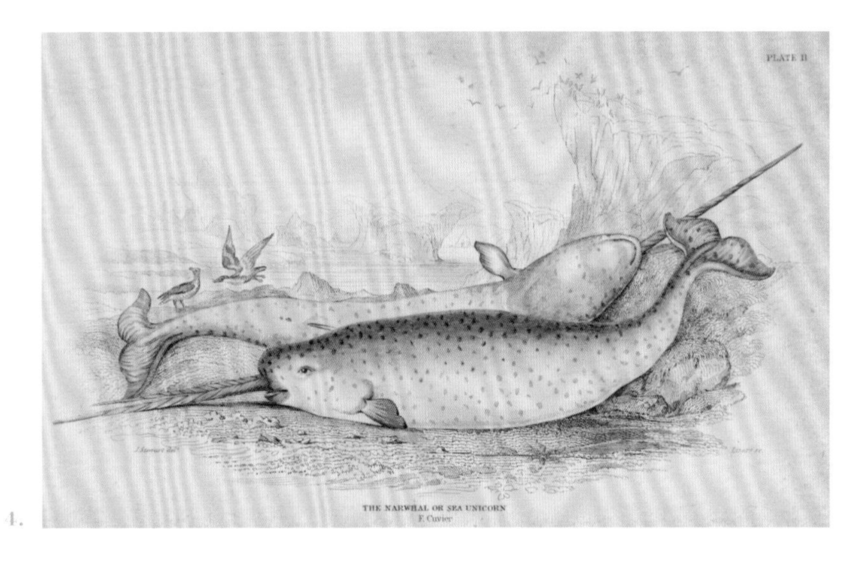

4.

3. Flying fish (Family Exocoetidae) leave the water powered by an elongate lower caudal fin lobe that rapidly sculls the sea surface and then spread their large pectoral fins for lift. In this way they can glide for distances of 50 meters (c. 165 feet) and more.

4. The Narwal's (*Monodon monoceros*) distinctive tusk, a remarkably elongated canine tooth, projects from the left side of the upper jaw through the lip and twists into a left-handed helix. Only male narwals have a tusk, and females look very similar to their close relatives, the beluga whales.

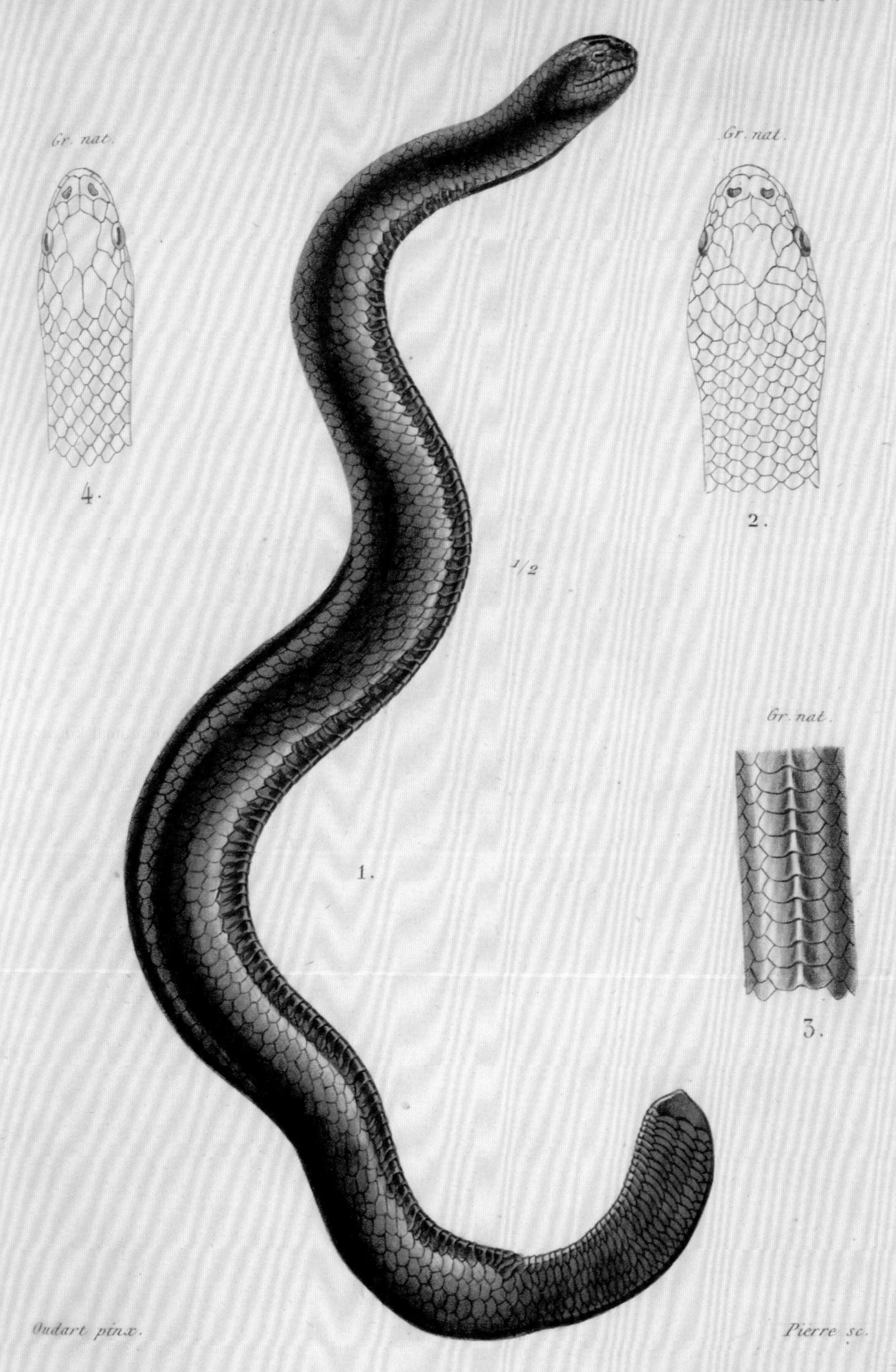

1. Aipysure fuligineux. 2. La tête vue en dessus. 3. Portion du tronc du même vue en dessous. 4. Tête de l'Aipysure lisse vue en dessus.

Signifiers of the Sea Snake

Authors
Constant Duméril
(1774–1860)
Gabriel Bibron
(1806–1848)
Auguste Duméril
(1812–1870)

Title
Erpétologie générale, ou histoire naturelle complète des reptiles

(General herpetology, or a complete natural history of reptiles)

Imprint
Paris: Roret, 1834–1854

André Marie Constant Duméril was a zoological savant of France's golden age of comparative anatomy, in a time of great geographical discovery and intellectual ferment. He was born in 1774 in Amiens in northern France and studied medicine in nearby Rouen, graduating at nineteen years of age. Despite his youth, Duméril served as provost of the Rouen Medical School until leaving for Paris in 1799, where he joined the faculty of medicine as chief of anatomical works. During this time, he came under the influence of his famous contemporary Baron Cuvier, and assisted him in the compilation of the influential *Leçons d'anatomie comparée*. And, when Baron Cuvier was appointed inspector of public instruction and occupied with postrevolutionary education reform, Duméril replaced him as instructor of natural history at the Panthéon (Sorbonne).

In 1803, on Baron Cuvier's recommendation, Duméril assisted and then succeeded his illustrious predecessor the comte de Lacépède (1756–1825) as professor of herpetology (the study of amphibians and reptiles) and ichthyology at the Natural History Museum in Paris, where he was to work for the rest of his career.

Duméril published broadly, with works covering the entire animal kingdom. In 1804 he published *Traité d'histoire naturelle*, dedicated to Baron Cuvier, and the following year saw the publication of the first edition of his celebrated *Zoologie analytique*. Despite the importance of these large-scale classificatory works, Duméril is best known for his detailed anatomical and herpetological studies; among these, the remarkable *Erpétologie générale, ou histoire naturelle complète des reptiles* stands as his most enduring legacy. This ambitious multivolume compendium was one of very few attempts ever made to cover all known species of reptiles and amphibians and remains an invaluable reference for students of herpetology. Published between 1834 and 1854, the work was prepared in collaboration with his trusted assistant, Gabriel Bibron (1805–1848), who had joined Duméril in 1832 after returning from Napoleon's famous military and scientific expeditions in the Greek Peloponnese.

Together they worked on the *Erpétologie générale* until 1845 when Bibron contracted tuberculosis, forcing him to leave Duméril for the retreat of Saint-Alban-les-Eaux near the Swiss border. Sadly, he never returned to Paris and died in Saint-Alban at the age of only forty-two. Bibron's death was a great personal loss for Duméril, but with the assistance of his son, Auguste Henri André Duméril (1812–1870), who by this time was an associate professor of comparative physiology at the University of Paris, he continued work on the compendium. In 1854 the ninth and final volume of *Erpétologie générale* was published along with a beautifully rendered atlas containing 108 plates.

The younger Duméril worked on completion of the seventh (in two parts) and the final ninth volume. It is in the second part of the seventh volume that the

1. The olive-brown sea snake (*Aipysurus laevis*) is the most common sea snake found on eastern Australian reefs. Like other sea snakes, it is highly venomous but generally docile, although it can become aggressive when breeding.

Dumérils discussed a curious sea snake, which they named the "sooty aipysure" (*Aipysurus fuligineux*), and illustrated it in the accompanying plate. Their description of the species was based on a single specimen collected by Maurice Arnoux, ship's surgeon on the French corvette *Le Rhin* during an 1842–1846 expedition to the Pacific Ocean. According to the Dumérils, Arnoux's specimen, collected off New Caledonia, differed from the two other sea snakes in the museum's collections of the time. One of these had been sent to the museum from Baudin's expedition (1800–1804) to Australia (see page 41) and named the "smooth aipysure" (*Aipysurus laevis*) by Lacépède, in 1804. Actually, the differences noted—which were mainly of coloration and in some details of the scales and teeth—are now considered to represent the kind of individual variation often exhibited among members of the same species, and Duméril's *Aipysurus fuligineux* is a synonym (see page 68) of the widespread, Indo-Pacific olive-brown sea snake (*Aipysurus laevis*).

While the great majority of the world's thirty-four hundred or so snake species live exclusively on land, a little more than sixty members of the Family Elapidae (to which cobras belong) are completely marine adapted and, like petrels among birds (see page 152), can no longer function on land. These are the sea snakes of the Subfamily Hydrophiinae, all of which live in the warm waters of the Indo-Pacific Ocean.

2.

Sea snakes exhibit numerous anatomical and behavioral adaptions to life at sea and are easily recognized by their flattened, paddlelike tails and laterally compressed bodies, features that greatly enhance their swimming abilities. Most species have reduced belly scales, and some have those scales modified into a sharp keel, as is so clearly illustrated in Duméril's plate of *Aipysurus*. In both cases, modification of the belly scales aids in swimming but renders these snakes virtually helpless on land.

They spend their entire lives at sea, where they hunt on coral reefs or among mangroves, feeding mainly on small fish, crustaceans, and soft-bodied mollusks. Most sea snakes also reproduce at sea; gestation is long, often up to nine months, and they give birth to well-developed young capable of immediately fending for themselves. Nearly all species are found inshore, often on and around coral reefs, where their short blunt heads aid in burrowing into crevices for food or protection from predators.

Elapid snakes are highly venomous and many sea snakes are among the most venomous of all. By way of comparison, the inland taipan (*Oxyuranus microlepidotus*) is one of the most venomous of land snakes, with one bite possessing enough venom to kill a hundred adult humans. But the bite of the faint-banded sea snake (*Hydrophis belcheri*) is estimated to be one hundred times more deadly. Happily, unlike their terrestrial cousins, sea snakes are not aggressive animals and most will only bite if highly stressed, as when accidentally caught in fishing nets, and even then less than one in four bites will release venom. Additionally, sea snakes, which are proteroglyphous, have "forward grooved" fangs rarely more than 0.5 centimeters (0.2 inches) in length, making the risk of envenomation of humans quite low.

2. The Dumerils included this frontispiece, a portrait of Gabriel Bibron by Marie Firmin Bocourt, in the last volume of the *Erpétologie générale* as tribute to his great contribution to the work.

3. Most elapids, like this broad-headed snake (*Holocephalus bungaroides*), are terrestrial and lack the modified ventral scutes of their sea snake relatives. Dumeril's artist, Paul Louis Oudart, has artfully arranged this specimen to reveal its ventral surface.

4. Here, the short, forward-grooved maxillary teeth of a proteroglyphous sea snake *Hydrophis curtus* (above) is contrasted with the highly effective solenoglyphous, or pipe-grooved, venom delivery system of the South American pit viper *Crotalus durissus* (center and below).

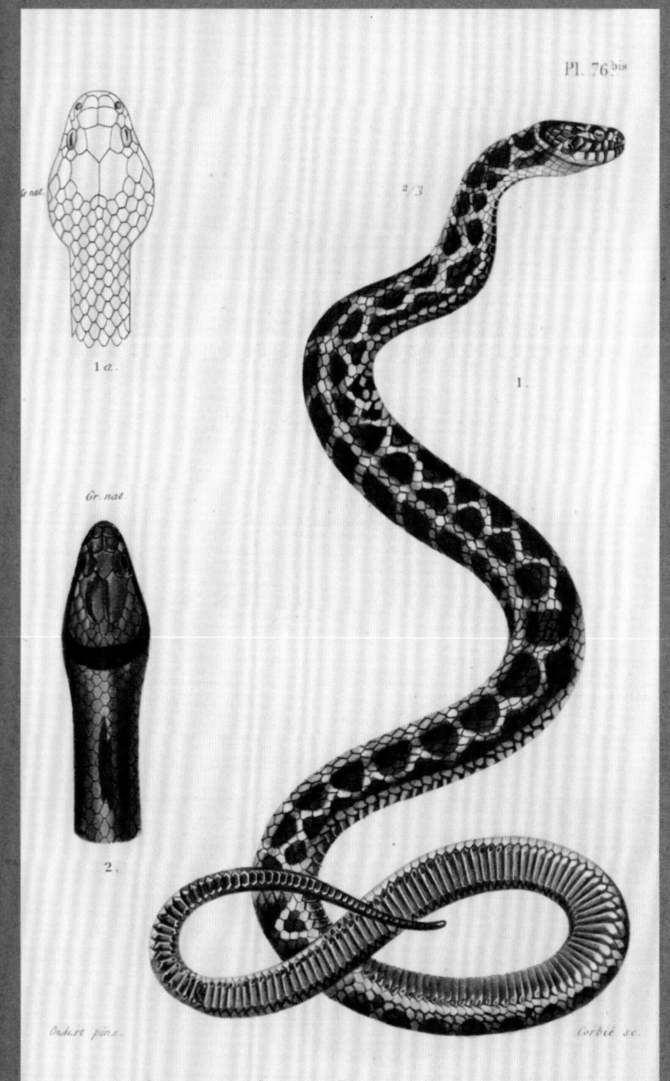

1. Alecto panachée. 1 a. Tête du même vue en dessus.
2. Tête de l'Alecto couronnée.

3.

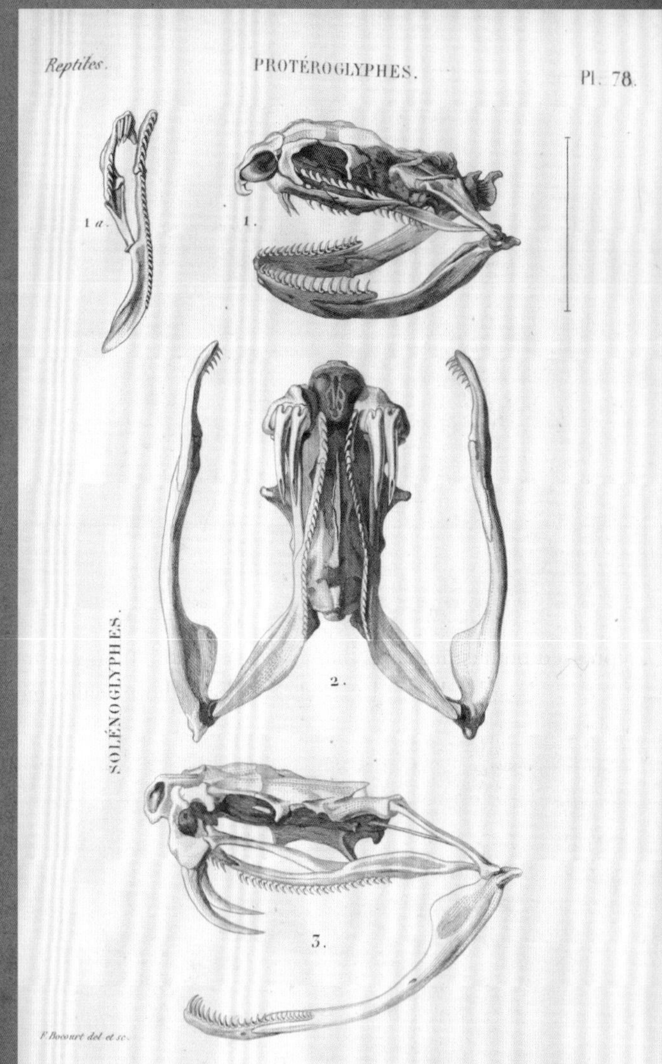

1, Hydrophis pelamidoïde; 1 a, Portion droite de la machoire supérieure; 2, Crotale durisse;
3, La même de profil.

4.

D'Orbigny's Stunning Starfish and Sea Hare

Author
Alcide Dessalines d'Orbigny (1802–1857)

Title
Mollusques, échinodermes, foraminifères et polypiers, recueillis aux îles Canaries par MM. Webb et Berthelot. (Histoire naturelle des îles Canaries, par mm P. Barker-Webb et Sabin Berthelot . . . t. 2, 2e ptie.)

(Mollusks, echinoderms, foraminiferans and polyps, collected from the Canary Islands by Messrs. Webb and Berthelot. [In Webb and Berthelot's Natural history of the Canary Islands, v. 2, pt. 2])

Imprint
Paris: Béthune, 1839

The death of his father in 1815 left the aristocratic British botanist and geologist Philip Barker Webb (1793–1854) heir to a fortune, freeing him to pursue scientific interests and explorations as he wished. He traveled extensively through Europe, and, after a series of forays around the Mediterranean and into North Africa, decided to visit Brazil. En route in September of 1828, Webb arrived on the island of Tenerife, the largest of the volcanic Canary Islands archipelago, isolated some 100 kilometers (c. 62 miles) off the coast of Morocco. It had been Webb's intention to stay only a short while but, before he was due to depart, he met Savin Berthelot (1794–1880), a young Frenchman of his own age who shared his enthusiasm for natural history and exploration. Berthelot had been living on Tenerife for the previous eight years and, together with Pierre Alexandre Auber (1784–1843), had founded a local school and managed a small botanical garden at Orotava on the northern coast of the island. Webb and Berthelot spent the next two years exploring the archipelago, making large zoological and botanical collections, studying the island's geology and geography, and learning about its indigenous peoples and their customs. Webb and Berthelot were exhaustive in their campaign, neglecting no physical, biological, or anthropological observation that would be needed to compile a comprehensive natural history of the island.

In April 1830, their work was done and they set sail for France, but at that time, Paris was entering the throes of the July Revolution and they settled instead in Geneva. By 1833 France had stabilized and they relocated to Paris and, with Webb's fortune, assembled a library and began preparing his impressive herbarium—a private collection said to be second in all of France only to that of Benjamin Delessert (see page 71).

In the following years, they worked together preparing materials for the *Histoire naturelle des îles Canaries*, which was to become a monumental compendium appearing in nine parts, assembled in three volumes, and published over a span of fourteen years between 1836 and 1850. Webb prepared the largest part of the *Histoire*, describing most of the geology and botany as well as a treatment of the island's small mammal fauna, while Berthelot contributed a history of the island's indigenous peoples, providing a comprehensive ethnography and detailed archeological and geographical observations. During the years of preparation of the *Histoire*, relations between Webb and Berthelot had become increasingly strained, and in 1846 Berthelot returned to Tenerife, where he remained until his death in 1880.

Traveling between his country estate in southern England and his residence in Paris, Webb continued working on botanical studies for the *Histoire*, but for the zoological treatises, he had recruited the services of many of France's most notable

1. While most starfish have only five arms, some, like this blue spiny starfish (*Coscinasterias tenuispina*), have more. Individuals of this species have from six to twelve arms.

specialists, one of whom was the famed explorer and naturalist Alcide Charles Victor Marie Dessalines d'Orbigny. D'Orbigny, based at the Paris Natural History Museum, had recently returned from his epic travels in South America and was working on the report of his findings, ultimately to be published in nine volumes as *Voyage dans l'Amérique méridionale*, a work considered by Charles Darwin to be "one of the great monuments of science in the nineteenth century."

D'Orbigny, the elder brother of Charles Henry Dessalines d'Orbigny (see page 103) and a favored protégé of the influential nineteenth century doyen of comparative anatomy, Baron Cuvier, was himself to become France's most revered paleontologist and founder of the science of biostratigraphy. But at the time he was approached by Webb, he was renowned as an accomplished zoologist who since his youth had studied marine invertebrates and was widely respected for groundbreaking work on marine foraminiferans (shelled amoeboid protists), which provided the foundation for the discipline of micropaleontology.

D'Orbigny studied the numerous invertebrate specimens collected by Webb and Berthelot and produced an extensive report published in 1839 as part of the second volume of the *Histoire* under the title *Mollusques, échinodermes, foraminifères et polypiers, recueillis aux îles Canaries par MM. Webb et Berthelot.* In the work, d'Orbigny described numerous new species, including the beautiful starfish depicted in the images that accompany this essay, one of which he named *Stellonia webbiana*, in honor of Webb.

The *Histoire* was richly illustrated with over four hundred plates of figures and illustrations, and Webb's wealth allowed him to commission many of the best natural history and landscape artists of the time. D'Orbigny's treatise was no exception and is accompanied by beautifully rendered illustrations, some of which, like the depiction of a spotted sea hare, were drawn by d'Orbigny himself.

Starfish are highly successful echinoderms belonging to the large Class Asteroidea. About sixteen hundred species live in the world's oceans, where they can be found from the intertidal zone to the ocean abyss, even thriving under the Arctic ice. Starfish, like many of their echinoderm relatives, such as the brittle stars (see page 26), are characteristically star shaped, with five (or multiples of five) arms radiating from a central disk. Like sea urchins (see page 124), starfish are able to move around using rows of tube feet that run alongside grooves on their undersurfaces.

Most starfish are voracious predators and can eat a wide range of foods including bivalve mollusks, which they encircle in their arms and grasp with their tube feet. Holding fast, the starfish slowly pries open the mollusk by tiring out its adductor muscle, which the mollusk uses to keep its shell shut tight. Once the shell is open a little way, the starfish is able to extrude its stomach into the opening and digest the mollusk's soft tissues.

Many starfish have a remarkable ability to regenerate lost or damaged arms and some, such as the one d'Orbigny named in Webb's honor, are able to regenerate a complete new disk from a single arm.

Ironically, many local fishermen consider the species a pest because it eats valued mussels and oysters and, when caught in fishing nets, is torn into pieces and thrown back into the water—only to regenerate into more individuals!

2. D'Orbigny described a number of starfish from the collections of Webb and Berthelot, among which were these beautiful specimens of *Astropecten aranciacus* (above) and *Narcissia canariensis* (below).

3. D'Orbigny's *Stellonia webbiana* (below), named in honor of its collector Philip Barker Webb, has since been synonymized with a species described some eighty years earlier by Linneaus, and is today known as *Marthasterias glacialis.*

4. The spotted sea hare (*Aplysia dactylomela*), a large aplysiid sea slug, regularly reaches 15 centimeters (c. 6 inches), with some individuals growing as large as 41 centimeters (c. 16 inches) in length. As a defense *Aplysia* exudes a cloud of purple ink, which appears to act as an irritant to potential predators.

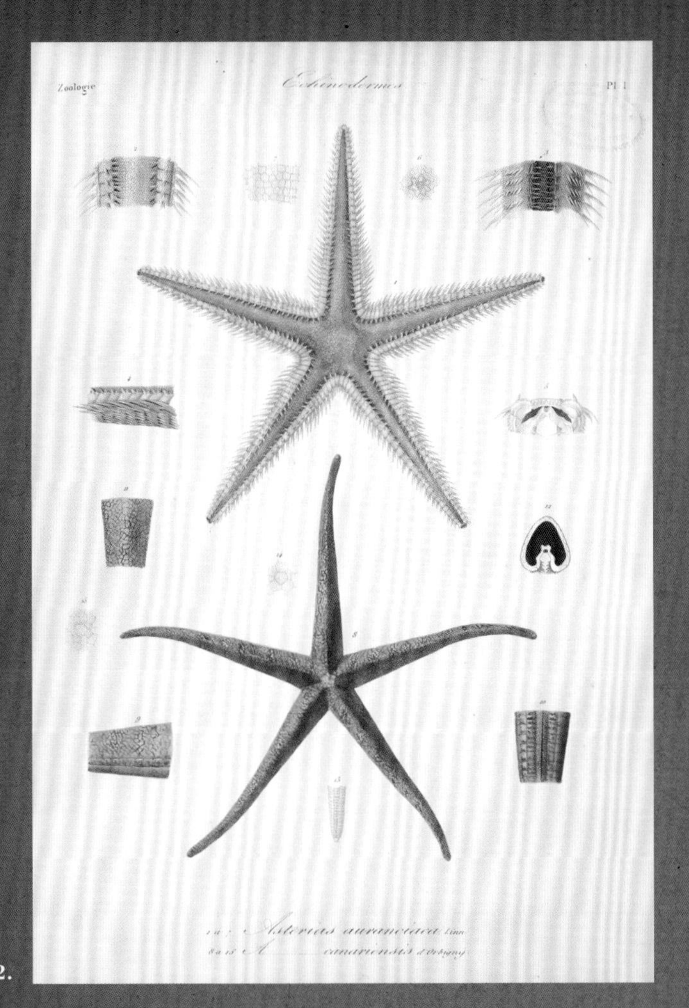
Zoologie
Echinodermes
Pl. 1
1 à 7 Asterias aurantiaca Linn.
8 à 15 A. canariensis d'Orbigny
2.

Zoologie
Echinodermes
Pl. 2
1 à 7 Ophidiaster ophidiana d'Orb.
8 à 13 Stellonia Webbiana d'Orb.
3.

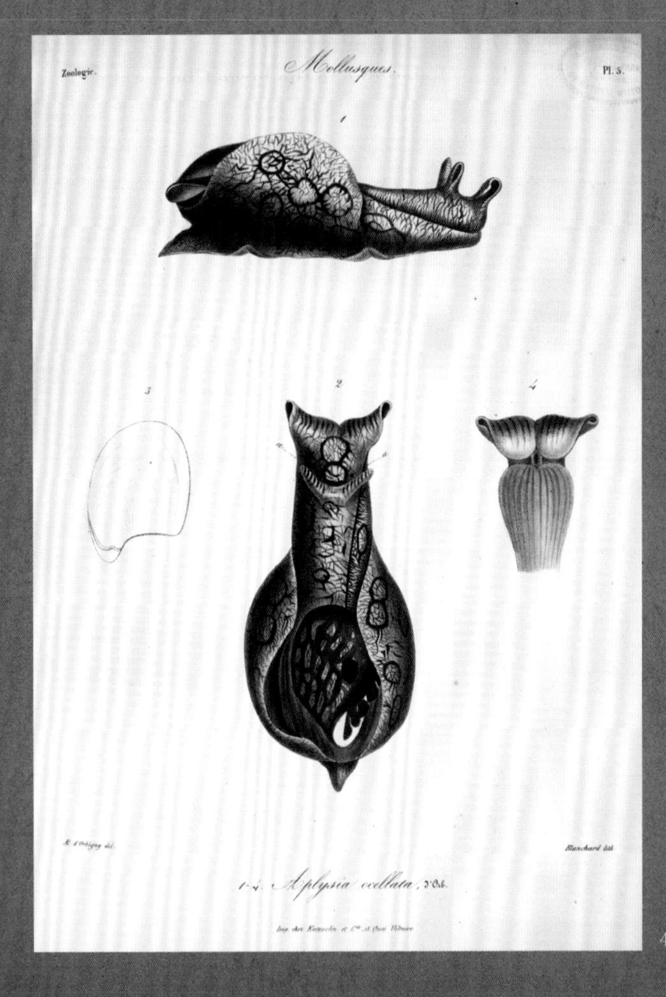
Zoologie
Mollusques
Pl. 3
1-4. Aplysia ocellata, d'Orb.
4.

Shortfin Mako: Swiftest Shark in the Sea

Author
Johannes Müller (1801–1858)

Title
Systematische Beschreibung der Plagiostomen

(Systematic description of the plagiostomes)

Imprint
Berlin: Veit und Comp., 1841

1. The beautifully streamlined and powerful shortfin mako (*Isurus oxyrinchus*) holds the speed record for the longest distance traveled—one individual swam over 2,100 kilometers (c. 1,305 miles) in thirty-seven days, averaging 58 kilometers (c. 36 miles) a day.

The German physician Johannes Peter Müller was a pioneering physiologist whose works helped to lay the foundation for the modern era of scientific study of human physiology. He was born into extremely modest circumstances in the Rhineland city of Koblenz, where his father was a cobbler. The intervention of one of his teachers saved him from having to enter into an apprenticeship in the leather trade, and he instead was able to attend a local college.

Initially aiming to enter the priesthood, Müller studied and excelled in the classics. By the age of eighteen, he had realized that the natural sciences held more appeal and in 1819, he enrolled as a medical student at the University of Bonn. After graduating in 1824, he assumed a lectureship in physiology and comparative anatomy at the medical school of the University of Bonn, and four years later he was appointed professor of physiology there. Müller became adept at microscopical studies and experimental physiology, and published a series of important papers on the neural control of human and animal visual and vocal systems, endocrinology, and reproductive anatomy, studies that established his international standing.

By 1833 Müller was chair of anatomy and physiology at Humboldt University in Berlin and between 1834 and 1840, his highly influential *Handbuch der Physiologie des Menschen* was published in two volumes, appearing in English translation as *Elements of Physiology* between 1837 and 1843. Müller's assistant at the time was Theodor Schwann (1810–1882). Schwann established the cell as the basic unit of animal structure, and soon after that, Müller began a series of studies of cell pathology. He worked particularly on cell pathologies associated with cancers, and his studies were to lay the foundation for the new field of pathological histology.

By the early 1840s, Müller refocused his researches on comparative anatomy and basic zoology, particularly of fishes and marine invertebrates. He led a series of expeditions to the Baltic and North Seas, and to the Adriatic and Mediterranean, where he and his students made large collections of specimens, many of which were to form the basis for his remarkably detailed anatomical and developmental studies. One of these resulted in an epochal treatise on the classification of sharks and rays, published in 1841 as *Systematische Beschreibung der Plagiostomen*. The work was produced in collaboration with Müller's former assistant, the eminent anatomist and physiologist Friedrich Gustav Jakob Henle (1809–1885).

Müller and Henle's monograph, which provided anatomical descriptions of 214 species of sharks and rays known at the time, was a seminal contribution and the classification that Müller produced placed many species into families and genera that are still recognized today. The work is accompanied by sixty beautifully rendered, hand-colored lithographic plates that illustrate not just the external appearance of each species, but also provide details of the shape and orientation of the jaws and dentition as well as the structure of the tiny toothlike (placoid) scales that encase each animal's body. Müller produced anatomical and classificatory studies on a wide range of organisms, from meticulous work on the anatomy and development of echinoderms to microscopic studies of marine foraminiferans and radiolarians (see page 120).

Müller was influential in many fields of medicine and zoology, and mentored numerous students and colleagues, many of whom—such as Hermann von Helmholtz (1821–1894), Theodor Schwann, Jakob Henle, and Ernst Haeckel (see page 119)—would become leaders in medical physiology and natural sciences themselves.

Müller died in Berlin, at the height of his fame and at the age of only fifty-seven. For much of his life, he had suffered debilitating bouts of depression, and it has been suggested that during one of these episodes, he may have taken his own life. Whatever the case, his death was greatly lamented by students, colleagues, and the scientific establishment alike. Müller had established Berlin as a leading center for medical research, and raised its museum to international stature. In 1899 a statue was erected in his honor in his hometown of Koblenz.

Sharks and rays and their close relatives the chimaeras (commonly called rabbit fishes) make up the vertebrate Class Chondrichthys (cartilaginous fishes), and Plagiostoma is an old term for the subgroup that includes just the sharks and rays (Elasmobranchii). Although they have a very long fossil record dating back over 420 million years, most modern elasmobranch groups began to appear in the fossil record only about 100 million years ago. The modern species bear little similarity to their ancient relatives. Of the thousand or so species alive today, about 560 are rays and a little more than 440 are sharks. While many of the groups of sharks and rays that Müller named are similar to those recognized today, ideas of how the different families of sharks are related to one another and how the rays fit in are topics of much current research. A clear picture has yet to emerge.

Müller and Henle provided a beautiful rendering of a shark they had named *Oxyrhina glauca*, but it turns out that Constatine Samuel Rafinesque (1783–1840) had described this same species over thirty years earlier, under the name *Isurus oxyrinchus*. Because Rafinesque's is the older name, it takes priority and *Oxyrhina glauca* is no longer used, and is considered a synonym. Regardless of its scientific name, the shortfin mako or blue pointer shark is one of the fastest animals in the oceans, able to maintain speeds of over 48 kilometers an hour (c. 30 mph) with bursts of upward of 72 kilometers an hour (c. 45 mph), and can perform incredibly acrobatic leaps, regularly reaching over 9 meters (c. 30 feet) in the air. It is a voracious predator, and able to undertake such high levels of activity because, like some other lamnid or mackerel sharks (including the great white), the mako is able to elevate its body temperature well above that of the surrounding water. It does this through a highly sophisticated counter-current heat exchange system, which can heat the brain, eyes, and swimming muscles up to 13 degrees Celsius (55 degrees Fahrenheit) above ambient temperature.

2. This shortnose guitarfish (*Zapteryx brevirostris*), like other members of the guitarfish Family Rhinobatidae, has a rather sharklike tail but is a true ray. Shown above the ray's head are its powerful jaws, which are used to crush bottom-dwelling crustaceans and polychaete worms.

3. The true identity of this shark, named *Carcharias* (*Prionodon*) *lamia* by Müller and Henle, is unclear. While some have suggested it is a great white (*Carcharodon carcharias*), details of its fins and teeth suggest that it is more likely a bull shark (*Carcharhinus leucas*).

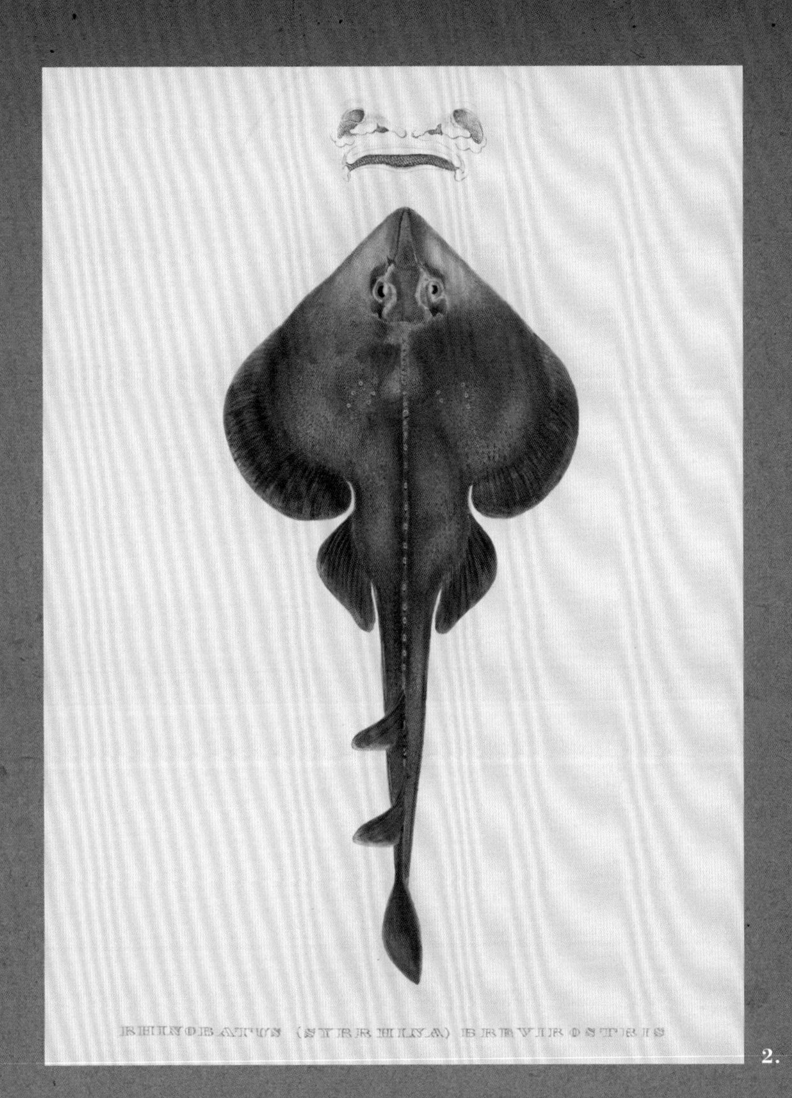

RHINOBATUS (SYRRHINA) BREVIROSTRIS

2.

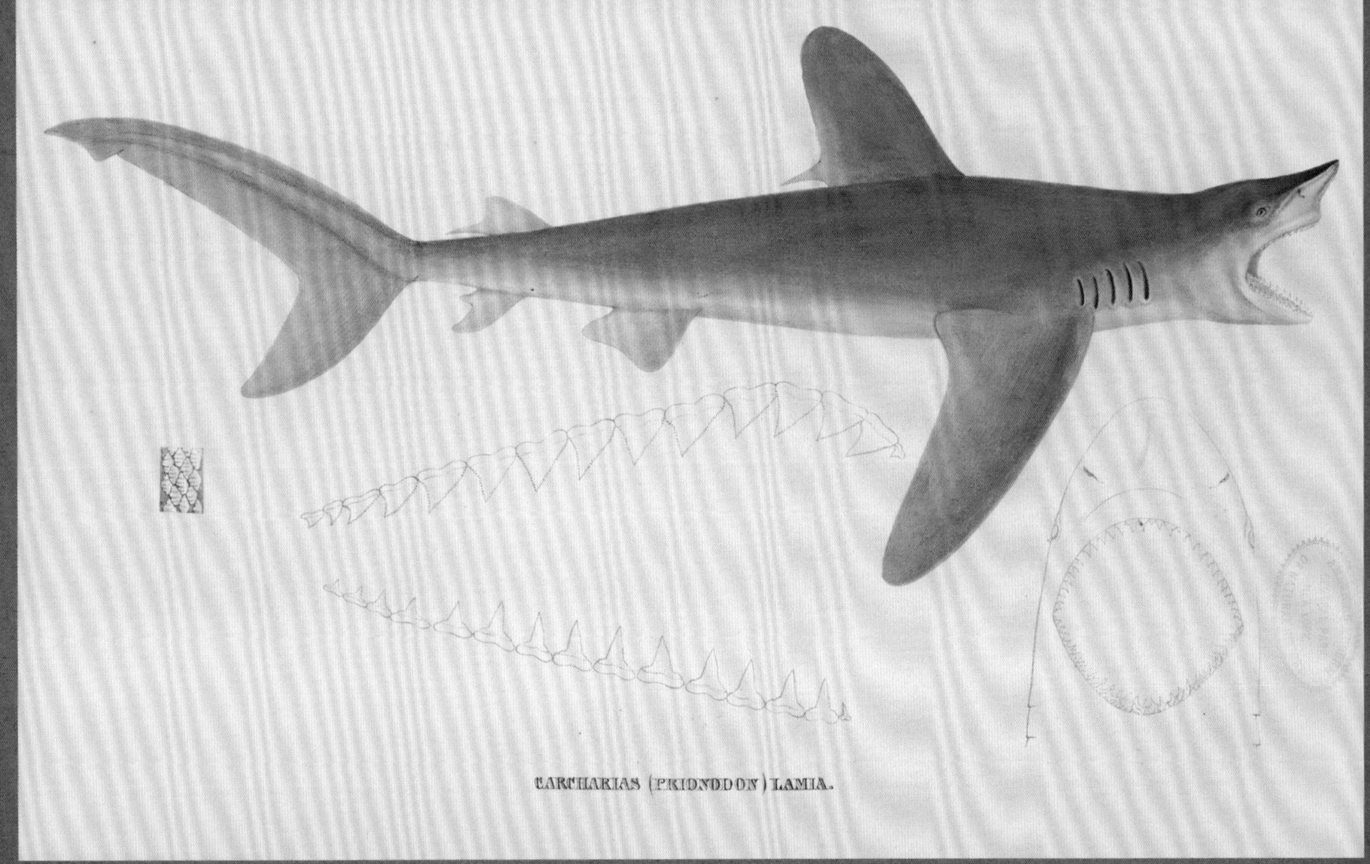

CARCHARIAS (PRIONODON) LAMIA.

3.

14 a
13 a
13 b
15 a
15 b
14 b
16 a
16 b

Smoking Cone Snails

Author
Jean-Charles Chenu
(1808–1879)

Title
Illustrations conchyliologiques ou description et figures de toutes les coquilles connues vivantes et fossiles, classées suivant le système de Lamark, modifié d'après les progrès de la science et comprenant les genres nouveaux et les espèces récemment découvertes

(Conchological illustrations or description and figures of all known living and fossil shells, classified according to the Lamarkian system, modified in accordance with the progress of science and including new genera and species recently discovered)

Imprint
Paris: Fortin, Masson, 1842–1854

The French physician Jean-Charles Chenu was born in the northern city of Metz, where he began medical studies, but he moved to Paris to continue his training. At the age of twenty-one, he entered military service as a field surgeon with the invading French armies during their campaign of conquest in Algeria. On returning to France, he completed medical training in Strasbourg, receiving a physician's license in 1833. Chenu was in southern France, in the department of Aude, when a particularly devastating outbreak of cholera struck the region. His successful treatment of the prefect of Aude, who had been stricken with the disease, gained him considerable repute and, during this time, he made the acquaintance of Gabriel and Benjamin Delessert.

Through his friendship with the Delessert brothers, Chenu obtained a government appointment as an inspector of the renowned mineral springs in the wealthy Paris neighborhood of Passy, where the Delesserts kept a family home. The family matriarch, Madame Delessert, was a close friend of Benjamin Franklin's during his years in Paris, and Franklin had enjoyed those same springs that Chenu was now charged with monitoring. While in Passy, Chenu published an important report on the treatment of cholera, and undertook a series of studies on the medicinal properties of Passy's iron-rich mineral waters.

Jules Paul Benjamin Delessert (1773–1847), Chenu's benefactor and the eldest of the Delessert brothers, was an important figure in Parisian society. He was a successful industrialist, inventor, and influential financier, and was also a keen amateur botanist and avid collector, whose vast fortunes had enabled him to amass spectacular personal collections of natural objects. Delessert had assembled an enormous private collection of shells from around the globe, said to contain over one hundred thousand specimens and to include many purchased for tremendous sums of money. One reputedly cost him six thousand francs—a truly staggering sum at that time. And it was to Chenu that he entrusted the curation and conservation of his private collections.

Between 1842 and 1853, Chenu published the first of his natural history studies, entitled *Illustrations conchyliologiques ou description et figures de toutes les coquilles connues vivantes et fossiles, classées suivant le système de Lamark*, and dedicated to Delessert "as a tribute in recognition of the profound respect of the author." The work was based in large part on Delessert's collection and appeared in four volumes, sumptuously accompanied by over 480 beautifully hand-colored, folio-size copper engraved plates.

Various prominent artists and engravers of the day contributed to this magnificent compendium, and among them was the famed Jean-Gabriel Prêtre (1800–1840),

1. The Glory of the Sea Cone (*Conus gloriamaris*) was for nearly two centuries considered the rarest shell in the world. In 1969 its habitat was discovered, and as more individuals entered the marketplace prices plummeted.

who meticulously illustrated the intricately patterned cone snail shells accompanying this essay. Chenu provided a review of all "shells" available to him, which is to say he rather unusually did not restrict his study to the shells of mollusks as did most conchologists (see page 108) of the time, but instead included a range of other marine organisms, such as tube-forming marine polychaete worms (see page 112) and barnacles (see page 85).

Many publications followed this landmark conchological study, and perhaps his most notable contribution was the ambitious *Encyclopédie d'histoire naturelle ou Traité complet de cette science d'après les travaux des naturalistes les plus éminents de toutes les époques*, a twenty-two-volume compendium published between 1850 and 1861, with contributions by many leading French naturalists of the day.

In 1868 Chenu retired from government service, but he continued to be actively involved with military medicine through a charitable organization he founded to care for wounded soldiers. He played an active role in the ambulance corps of the charity during the disastrous Franco-Prussian War of 1870. Chenu died in Paris in 1879 at the age of seventy-one.

The intricately patterned and colored shells of cone snails made them a favorite of collectors through the ages, and Delessert was no exception. Chenu depicts numerous specimens representing color and pattern variants of about sixteen species of these stunning marine gastropod mollusks.

Cone snails are typically found in tropical seas, burrowing in sand or among rock and coral rubble on or around reefs. All are highly predatory, feeding mainly on marine worms and small fish. These slow-moving snails utilize an efficient prey capture mechanism involving a venomous barb to harpoon their fast-moving prey. One radular tooth is enlarged, barbed, and hollow (in most gastropods, the radula has only rows of small teeth), and when a prey is detected, the snail rapidly ejects its long proboscis bearing the barbed tooth at the tip. The tooth is connected to a venom gland, and the snail injects neurotoxic venom that immobilizes the prey, which is drawn into its mouth and digested. After each envenomation, a new barbed tooth developed in the radula is recruited for the next attack.

Cone snails produce very complex venoms (conotoxins) containing hundreds of neurotoxic peptides. Some larger species, which specialize in capturing fishes, are hazardous to humans as their tooth can readily penetrate skin, even through neoprene, with occasionally fatal consequences. The Indo-Pacific *Conus geographus* is among a group of highly toxic cone snails and gained its common name, the "cigarette snail," from the suggestion that between the time you are stung and when you die, you'd have about enough time to smoke a cigarette—an exaggeration, but nonetheless there are documented cases of human death caused by cone snail envenomation.

On the positive side, conotoxins show considerable promise as pharmaceuticals, particularly in the field of pain management. In 2004, the U.S. Food and Drug Administration approved the first painkiller derived from a conotoxin.

2. Many species of polychaetes, including these strange spaghetti worms (Family Terebellidae), build tubes within which the soft-bodied animals are protected while they filter surrounding water for food, using highly modified tentacles.

3. These convoluted tubes are the shells of worm snails (*Vermetus*). These strange gastropod mollusks produce irregular, elongate, tubular shells that are cemented to rocks and other shells.

4. Of the many hundreds of specimens in Delessert's remarkable collection, it was undoubtedly the intricately patterned, and at the time extremely rare, cone shells that were considered the most prized of all.

Pl. 5
G. TEREBELLA. Linné.
T. Olfersii
T. Tondi. id.
T. Meckelii. id.
T. Nesidensis.
T. Flexuosa. id.
T. Neapolitana. id.
G. AMPHITRITE. Lamarck
A. Magnifica.
A. Mulleri.

2.

ILLUSTRATIONS CONCHYLIOLOGIQUES.
G. VERMETUS. Adanson.
V. Dentiferus.
V. Margaritaceus.
V. Sipho.
V. Turonius.
V. Gigas. Philippi.
V. Peronii.

3.

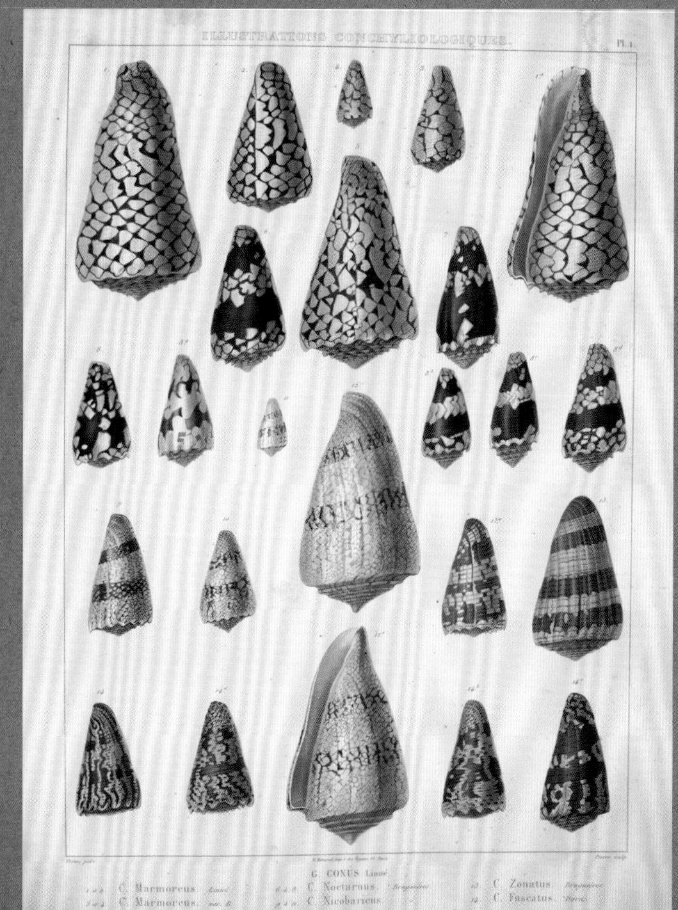
ILLUSTRATIONS CONCHYLIOLOGIQUES.
Pl. 1
G. CONUS Linné.
C. Marmoreus.
C. Marmoreus.
C. Bandanus.
C. Nocturnus.
C. Nicobaricus.
C. Araneosus.
C. Zonatus.
C. Fuscatus.

4.

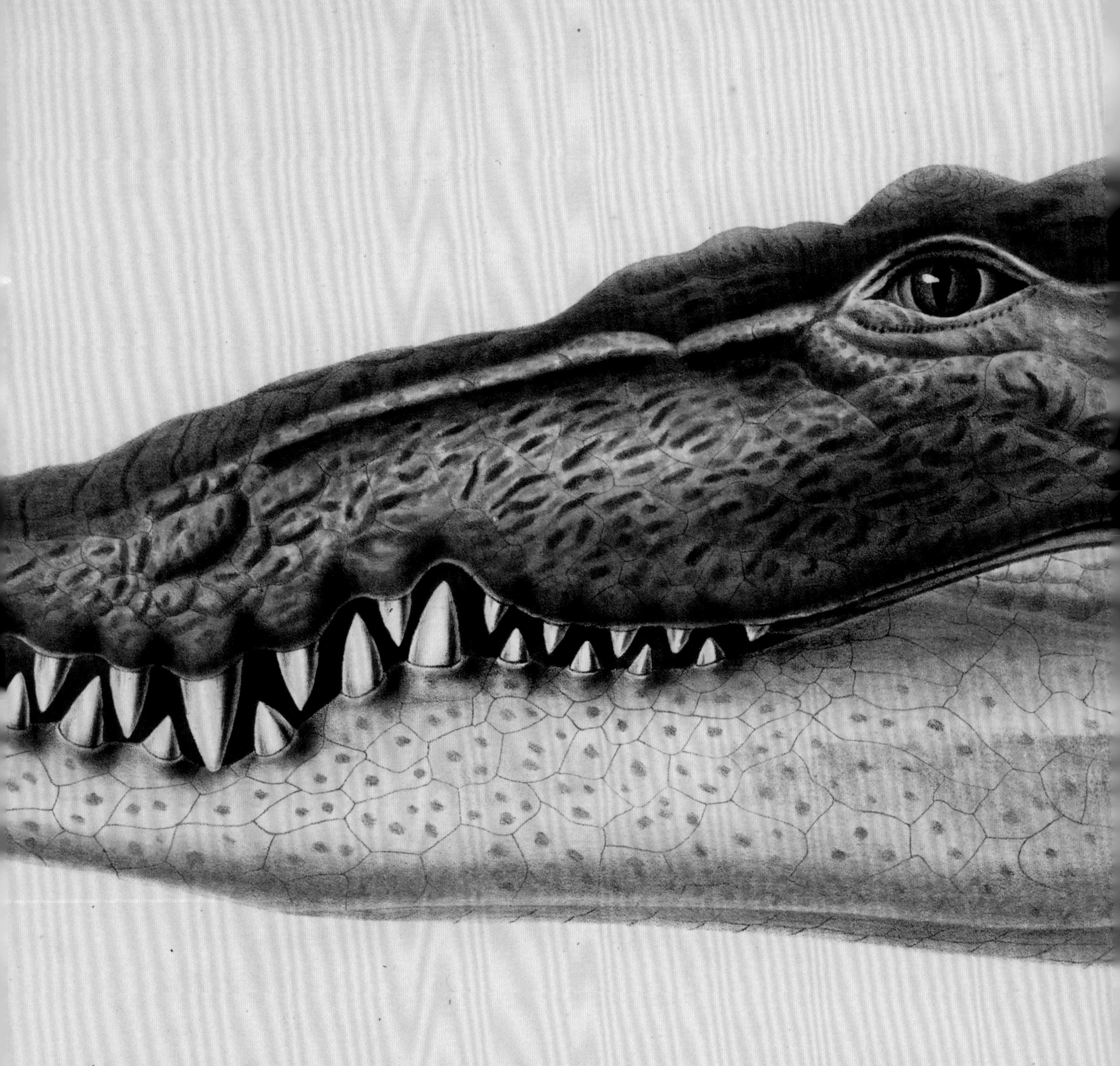

Crocodilus biporcatus.

Schlegel's Guide to Amphibians

Author
Hermann Schlegel
(1804–1884)

Title
Abbildungen neuer oder unvollständig bekannter Amphibien, nach den Natur oder dem Leben entworfen

(Drawings of new or imperfectly known amphibians drawn from life in nature)

Imprint
Düsseldorf: Arnz & Comp., 1844

Figure 4
Author: Franz Steindachner (1834–1919)
Title: *Die schlangen und eidechsen der Galapagos-inseln*
Imprint: Wien: K.K. Zoologisch-botanischen gesellschaft, 1876

1. Saltwater crocodiles (*Crocodylus porosus*) have the strongest bite of any living animal, and as adults can easily crush the skull of a full grown bovid. The largest specimen recorded was a male, shot in the Bay of Bengal in 1840, and measured 10 meters (c. 33 feet) long.

Hermann Schlegel was born in the town of Altenburg in northern Germany. Despite an early interest in natural history, particularly the study of birds, Schlegel began an apprenticeship at his father's brass foundry. Unsatisfied with the work, at the age of twenty he decided to pursue his interests in natural history and left Altenburg for Vienna. While in Vienna, Schlegel met many of the leading naturalists of the day and, with the help of these influential acquaintances, was introduced to Joseph Natterer (1787–1843) in consideration for a position at the Viennese Natural History Museum.

Apparently Schlegel made a positive impression, as one year later the museum's director recommended him to serve as assistant to the eminent Dutch aristocrat and naturalist Coenraad Jacob Temminck (1778–1858), the first director of the Dutch Natural History Museum in Leiden. Schlegel moved to Leiden and began his career under Temminck's tutelage, soon becoming his most notable student. Thirty-three years later, after Temminck's death, Schlegel succeeded him as director of the Leiden museum, where he remained until his death in 1884.

It was originally intended that Schlegel would travel to Batavia (present-day Jakarta), at the time the center of Colonial Dutch East Indies. Unfortunately, circumstances in Leiden prevented him from leaving and it was at this time that he met the famed botanist and explorer Philipp Franz von Siebold (1796–1866), who had returned to Holland after working in Japan some years earlier (see page 107). The two established a firm friendship and both Schlegel and Temminck contributed to Siebold's magnificent four-volume opus, *Fauna Japonica*.

Although perhaps best known for his studies of birds, Schlegel had initially focused on amphibians and reptiles, and the work, *Abbildungen neuer oder unvollständig bekannter Amphibien, nach den Natur oder dem Leben*, was among the first that he published after settling in Leiden. This beautifully illustrated work was based mainly on drawings received from naturalists working in India under the auspices of the Dutch government. Schlegel believed that the absence of good illustrations of amphibians was an impediment to their study and somewhat optimistically felt that publication of a magnificent series of colored drawings would generate public interest in them. He cited the many works on birds and a much admired compendium on mammals published by Frédéric Cuvier (1773–1838) as examples of how good illustrations had stimulated interest in those subjects. With this goal in mind, Schlegel compiled the work, which includes fifty color plates assembled into an atlas that is accompanied by a short volume of text. Schlegel was concerned that publication of the illustrations should not be delayed, and chose to defer presenting formal descriptions of any new species until a later date.

Schlegel's stated motivation for the publication was to encourage interest in amphibians, and the title of the work eludes only to amphibians—yet many more reptiles than amphibians are discussed and illustrated in the work. Whether this was an oversight, and a provisional title mistakenly remained unchanged after the scope of the work broadened, is unknown. Also unknown are the names of the artists who supplied the vivid renderings, and Schlegel mentions only that they were received from different painters in India. Regardless, one of the most striking of the plates selected by Schlegel for inclusion in this work is the beautifully rendered head of a saltwater crocodile. The name used by Schlegel, *Crocodylus biporcatus*, is no longer considered valid, and the only oceangoing crocodile recognized today is *Crocodylus porosus*. It is this species that is so excellently represented in the accompanying plate.

It is worth noting that there are very few modern amphibians or reptiles living in the world's oceans. Of approximately seven thousand amphibian species, only a single frog (*Fejervarya cancrivora*) is able to tolerate immersion in salt water for a short period, and it spends most of its time at the ocean's edge in mangroves.

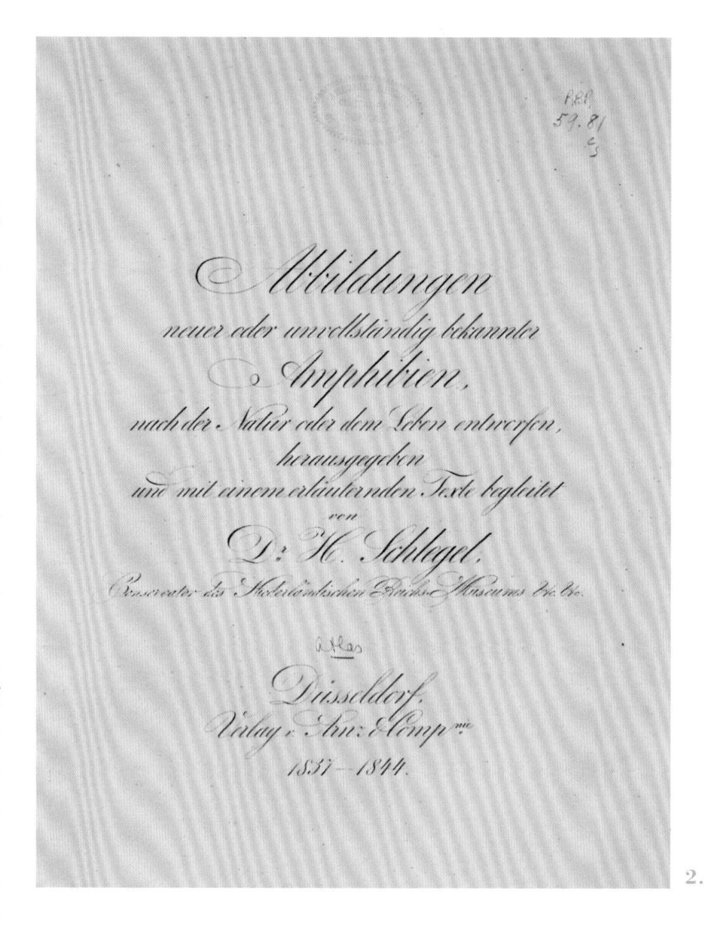
Abbildungen
neuer oder unvollständig bekannter
Amphibien,
nach der Natur oder dem Leben entworfen,
herausgegeben
und mit einem erläuternden Texte begleitet
von
Dr. H. Schlegel.
Conservator des Niederländischen Reichs-Museums &c. &c.
Atlas
Düsseldorf.
Verlag v. Arnz & Comp.
1837–1844.

2.

2. This curious, handwritten title page accompanies Schlegel's Atlas of drawings.

3. Despite the work's title, most of the animals featured by Schlegel are reptiles, and this plate depicts a species of sea snake found throughout the Indo-Pacific, the dwarf sea snake (*Hydrophis caerulescens*).

4. Darwin's famous marine iguana (*Amblyrhynchus cristatus*) is illustrated here in a plate published in 1876 by Franz Steindachner in a work on the snakes and lizards of the Galapagos Islands.

Among squamate reptiles (iguanas, lizards, and snakes), only sea snakes (see page 60) have adapted to marine habitats in any numbers, although a single species of iguana (*Amblyrhynchus cristatus*), so famously observed in the Galapagos by Charles Darwin (1809–1882), is also considered marine. And finally, of 260 species of testudines (turtles, tortoises, and terrapins), only seven species of sea turtles (see page 34) have a fully marine lifestyle.

The situation in the ancient oceans was very different when numerous archaic amphibians and reptilian groups—such as ichthyosaurs, pliosaurs, plesiosaurs, and mosasaurs—were dominant predators of the world's oceans.

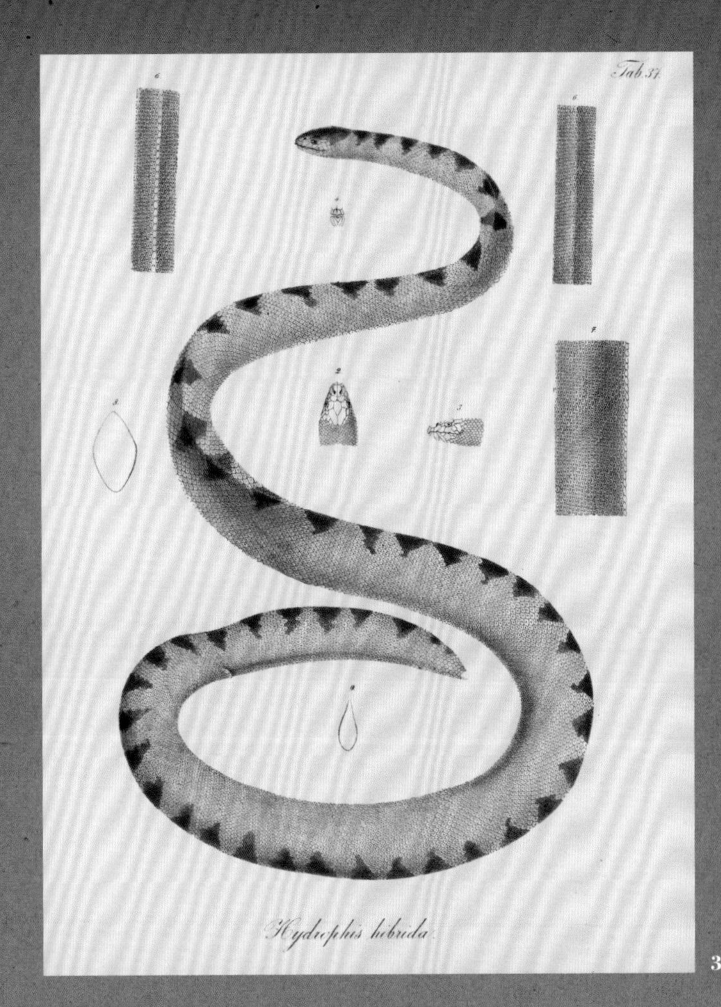
Tab. 37
Hydrophis hebrida
3.

F. Steindachner. Schlangen u. Eidechsen der Galapagos-Inseln.
Taf. III.
Festschrift d. k.k. zoolog. botan. Ges. in Wien 1876.
4.

W.H.H. delᵗ

Engᵈ by J. Petrison, Dubˡⁿ

The Gentleman Phycologist

Author
William Henry Harvey (1811–1866)

Title
A manual of the British marine algae: containing generic and specific descriptions of all the known British species of sea-weeds

Imprint
London: John Van Voorst, 1849

1. Brown algae are among the largest and fastest growing of seaweeds. Some, such as the winged kelp (*Alaria esculenta*) illustrated on the left and the dead man's rope (*Chorda filum*) on the right, are traditional foods around the eastern North Atlantic coasts.

The Irish botanist William Henry Harvey was born the youngest child of wealthy Quaker merchants in a small town on the Shannon estuary near Limerick. From early childhood, Harvey was captivated by the natural world, and by his early teens was an enthusiastic amateur phycologist (the study of algae). Much of his free time was spent roaming the Irish coastline and countryside in search of seaweeds and mosses, and as a young man he made the fortuitous discovery of the moss *Hookeria laetevirens*, growing near Killarney. This moss was known from the Mediterranean and its discovery in Ireland underscored an interesting geographical pattern that botanists at the time were puzzling over.

Sir William Jackson Hooker (1785–1865), the renowned botanist for whom the genus was named, was the popular Regius professor of botany at Glasgow University and, on learning of Harvey's botanical talents, began a correspondence with him and a friendship that was to last until the end of both their lives. Hooker mentored Harvey and in 1833 invited the promising young phycologist to write the algal sections in a new edition of his famous *British flora*, and for his *Botany of Beechy's voyage to the Pacific and Behring's Straits*, published in 1841. He introduced Harvey to many of the botanical elite of the day, men such as the brilliant botanical microscopist Robert Brown (1773–1858) and James Ebenezer Bicheno (1785–1851), who was to become the colonial secretary of Van Diemen's Land (Tasmania), as well as to Hooker's own son, one of Charles Darwin's closest confidants, Joseph Dalton Hooker (1817–1911).

In 1835 Harvey and his brother Joseph journeyed to the British Cape Colony (South Africa) to take up government positions, but Joseph's health declined rapidly and the brothers left for Ireland. Sadly, Joseph died on the voyage, but Harvey returned to Cape Town as treasurer-general. In Cape Town he began work, together with Otto Wilhelm Sonder (1812–1881), on the monumental *Flora capensis*. For this Harvey recruited the aid of numerous local naturalists and collectors, among them the remarkable Mary Elizabeth Barber (1818–1899), a pioneering woman naturalist. Barber lived at a time that was extremely hostile to the participation of women in science; indeed, in their initial correspondence, she withheld her gender from Harvey. Over many years of friendship, she helped him name and classify numerous species and sent him many hundreds of specimens, each with detailed notes. Barber also corresponded with Darwin, Joseph Hooker, and other notable scientists of the day who appreciated her talents and scientific contributions, despite her gender.

Harvey returned to Ireland in 1842 and two years later received an honorary doctorate from Trinity College, Dublin. He was appointed curator of botany at the university's Herbarium. During this time, he completed the second edition of his highly acclaimed *Manual of the British marine algae*, published in 1849 and accompanied by twenty-eight hand-colored lithographic plates. Highly unusual for the time, he dedicated the edition to another extraordinary woman, "Mrs. Griffiths of Torquay." This was Amelia Griffiths (1768–1858), an amateur phycologist who had an exceptional knowledge of the algae she collected and observed, and Harvey held her in the highest esteem. With characteristic generosity—and modesty—he wrote, "A lady whose long-continued researches have, more than those of any other observer in Britain, contributed to the present advanced state of marine botany" and that "this volume ... owes much of whatever value it may possess to her liberal donations of rare specimens, and her accurate observations upon them."

In 1848 Harvey became professor of botany of the Royal Dublin Society and, in 1856, he was appointed chair of botany at Trinity College, positions he held until his death from tuberculosis in 1866.

Harvey's contributions to phycology were enormous; he traveled the globe and published major works on the marine floras not just of Britain and the Cape, but also of the Americas, Australia, Tasmania and the southern oceans, the Arctic, and the Antarctic. Shortly after his death, Asa Gray (1810–1888), considered by many to be North America's greatest botanist, wrote this of Harvey: "Handsome in person, gentle and fascinating in manners, genial and warm-hearted but of very retiring disposition, simple in his tastes and unaffectedly devout, it is not surprising that he attracted friends wherever he went, so that his death will be sensibly felt on every continent and in the islands of the sea."

Seaweed is a term used to describe a large and diverse assemblage of photosynthetic, plantlike marine organisms, divided since the early nineteenth century into three main groups: brown (Phaeophyta), red (Rhodophyta), and green (Chlorophyta) algae. Although their colors can vary considerably, each group shares its own particular kind of photosynthetic pigment that gives it its predominant color. Despite many similarities, these algal groups are not at all closely related.

Brown algae, with about eighteen hundred species, including such massive organisms as the giant kelps, surprisingly are more closely related to tiny diatoms (unicellular phytoplankton) and oomycetes (microscopic pathogens such as those that cause diseases like potato blight) than they are to the other algae. Red algae, with about six thousand species, are very important primary producers in marine habitats, and include the so-called coralline algae, which secrete calcium carbonate and help build coral reefs. Finally, the green algae, with about four thousand species, are the only algae that are closely related to land plants and are included in the Kingdom Plantae. They utilize the same chlorophylls for photosynthesis as other plants, and also store their food as starch.

Seaweeds, regardless of their relationships, are a critically important part of the marine ecosystem and play a very similar role there as plants do on land.

Harvey's manual was a foundational early work on British algae, and the following four plates represent a small selection of the forms he figured:

2. Phaerophytes, or brown algae A) *Sargassum*, B) *Cytoseira*, C) *Halidrys*, D) *Fuscus*

3. Rhodophytes, or red algae A) *Rhodymenia*, B) *Sphaerococcus*, C) *Gracilaria*, D) *Hypnea.*

4. Phaerophytes, or brown algae A) *Cutleria*, B) *Haliseris*, C) *Padina*, D) *Zonaria*

5. Rhodophytes and Chlorophytes, or green algae A) *Porphyra* (red), B) *Ulva* (green), C) *Bangia* (red), D) *Enteromorpha* (green), E) *Ochlochaete* (green)

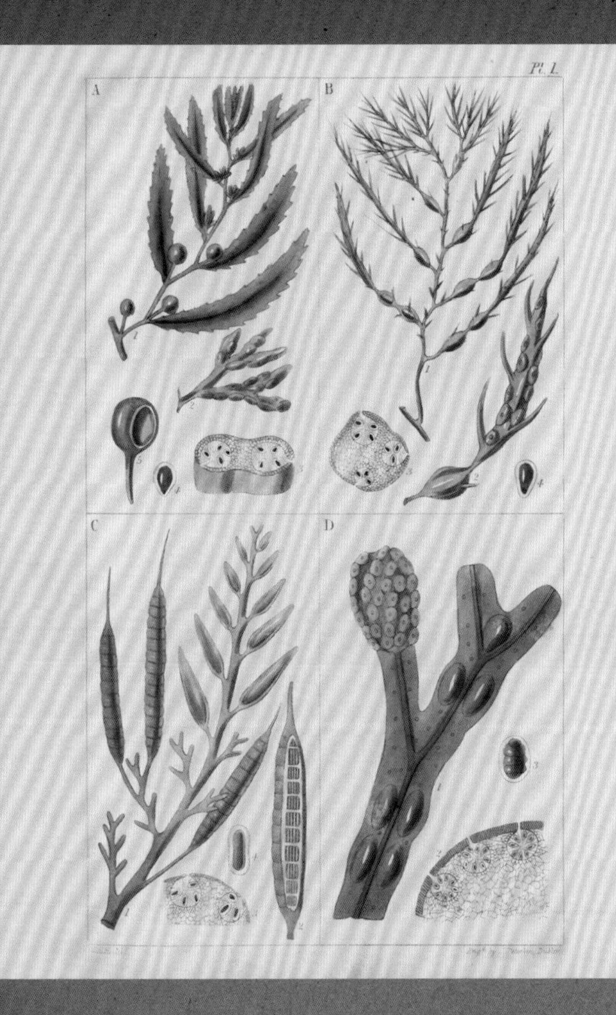

2.

3.

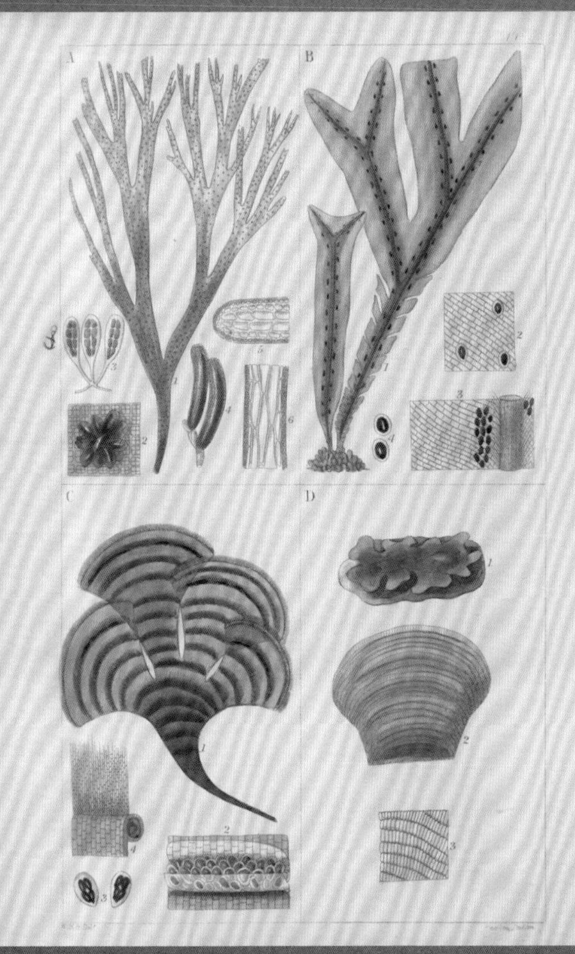

4.

5.

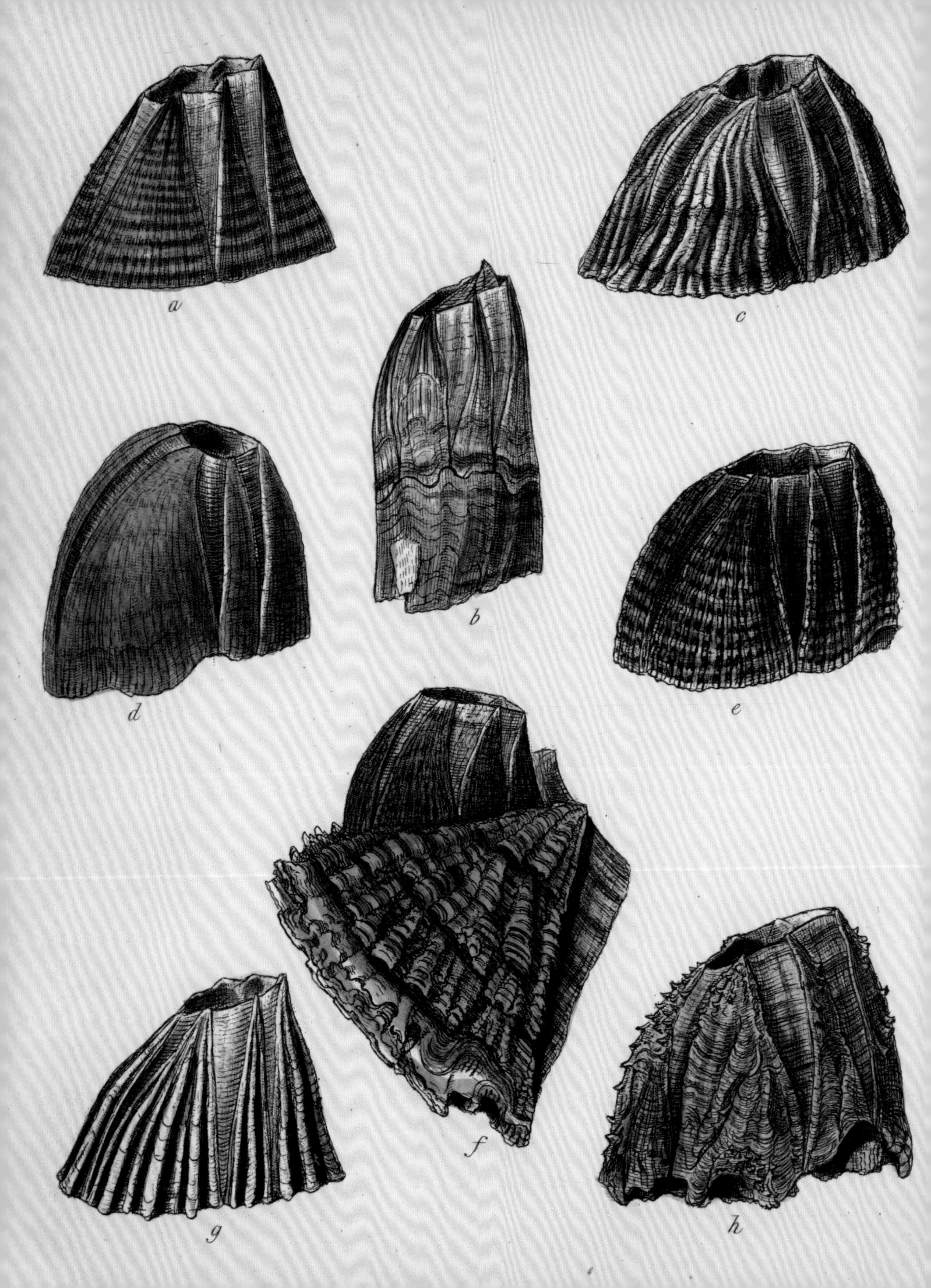
a
c
b
d
e
f
g
h

Darwin's "Beloved Barnacles"

Author
Charles Darwin
(1808–1882)

Title
A monograph on the sub-class Cirripedia, with figures of all the species

Imprint
London: Ray Society, 1851–1854

Figure 3
Author: Theodore Edward Cantor (1809–1860)
Title: *Observations upon Pelagic Serpents* (Trans. Zool. Soc. London, v.2)
Imprint: London: The Society, 1839

1. Darwin considered the large acorn barnacle *Megabalanus tintinnabulum* to represent just one highly variable species. This barnacle is thought to be of tropical origin, but it has been spread around the globe attached to ship hulls.

Charles Darwin is rightly celebrated for his epoch-making theory of evolution by natural selection, with its simple but iconoclastic premise that all life shares common ancestry. And, although Darwin is popularly associated with his famous Galapagos finches and giant tortoises, he also had a lifelong fascination with marine invertebrates. In 1825 the young Darwin had been sent by his father to study medicine in Edinburgh—and he hated it. Happily, he found the perfect distraction in the Plinian Society, formed to encourage the study of natural science among Edinburgh students. The Plinians gathered weekly for discussion and nature walks along the nearby Firth (estuary) of Forth.

Darwin's mentor, the zoologist Robert Edmond Grant (1793–1874), sparked his interest in marine invertebrates, teaching him to collect and rear specimens. In March of 1827, Darwin presented his first scientific paper to the Plinian Society, demonstrating that bryozoan larvae ("moss animals") use cilia for locomotion, and that the black spots often observed on oyster shells were in fact eggs of marine leeches. For the aspiring nineteen-year-old, these were noteworthy findings, but sadly Grant had trumped Darwin's debut by presenting these same findings three days earlier to a different audience. Understandably, Darwin felt betrayed, and his relationship with his former mentor was never to recover. After just two years in Edinburgh, Darwin left in disgust, not having the stomach for the medical profession or its professors.

On Darwin's return to the Mount, his family home near Shrewsbury, his increasingly exasperated father dispatched him to Cambridge, where from 1827 to 1831 he studied for a bachelor of arts degree—a prerequisite for entrance into the clergy. In truth, Darwin had as little interest in the priesthood as he had in medicine, and again it was to natural history he turned for solace, this time under the mentorship of the influential botanist and geologist John Stevens Henslow (1796–1861). After graduation, and on Henslow's recommendation, Darwin joined Robert FitzRoy, captain of the HMS *Beagle*, as a gentleman naturalist and companion, on a cartographic voyage around the world.

During the famous five-year voyage of the *Beagle*, Darwin's interests in marine biology were reignited. He made countless observations, took copious notes, and amassed huge collections, including many marine invertebrates, regularly sending specimens and screeds of observations back to Henslow in Cambridge. This time Darwin had a mentor worthy of his trust, one who helped bolster his standing in the scientific community by distributing summaries of his observations. Thanks to Henslow, on Darwin's return in 1836, his reputation was established, and he set about the monumental task of working up his notes and distributing his vast collections among experts for identification and description.

By 1842 Darwin published his groundbreaking study *On the structure and distribution of coral reefs* and, in the following year, the final volume of the massive *Zoology of the voyage of H.M.S. Beagle* was published. By this time, Darwin had begun to sketch out his ideas on the mutability of species and the mechanisms of natural selection—ideas he half-jokingly described as being tantamount to "confessing a murder"—and tentatively began to share them with close colleagues. In the midst of this intellectual activity and spiritual turmoil, Darwin began an eight-year study of something so seemingly banal that many historians have interpreted it as a monumental digression. This could not be further from the truth, for Darwin had returned to his early love—marine invertebrates—and it was the study of one group, the barnacles (Cirripedia), that would solidify his credentials as a taxonomic expert and provide him with empirical support for many of his evolutionary ideas.

Ten years earlier, on the coast of Peru, Darwin had found a conch shell riddled with holes. Characteristically, he was intrigued, and closer examination revealed what he thought may be some kind of burrowing barnacle, which he humorously described as an "ill-formed little monster" and named "Mr. Arthrobalanus." In 1846 he returned to that strange, burrowing barnacle with the intent of providing a description of this singularly "abnormal Cirripede from the shores of South America," but soon realized that to do so he would need to examine many more species. Corresponding with experts around the globe and borrowing countless specimens of living and fossil species, he began a meticulous study of what were to become his "beloved barnacles."

Study of these bizarre crustaceans proved a fortuitous choice, providing Darwin with confirmation for many of his evolutionary ideas—such as the loss of unnecessary structures (barnacles have no trace of the abdominal segments and swimming appendages of other crustaceans), and evidence that features inherited from a common ancestor can transform in anatomy and function (the typical walking limbs of crustaceans are modified into specialized feeding cirri in barnacles).

After eight years of sometimes frustrating but endlessly stimulating work, Darwin published his monumental four-part series, *A monograph on the sub-class Cirripedia, with figures of all the species.* These works were milestones of modern monographic science, establishing Darwin's credentials as a taxonomic expert and giving him the credibility he felt he needed to publicly expound his ideas on the origin of species. It really is no coincidence that the word "cirripede" or "cirripedes" appears twenty-six times throughout his masterwork *On the origin of species*, published in 1859, just five years after he completed his monumental barnacle studies.

2.

2. Darwin's studies included the stalked varieties, also known as goose barnacles. The later name derives from the curious medieval belief that the Barnacle Goose (*Branta leucopsis*) developed from these strange, stalked structures. Conveniently, when meat was forbidden during religious holidays, the geese, not being "born of flesh," could be eaten.

3. This plate, published by Theodore Cantor in 1839, shows stalked barnacles removed from the scales of a slender sea snake (*Hydrophis gracilis*). Sea snakes shed their skins more often than terrestrial snakes, probably as a means of removing the barnacles and the many other epibionts that attach to them.

4. Embryonic development fascinated Darwin, and his meticulous studies of the life cycle of barnacles with the extraordinary transformation of their nauplius larvae provided him further evidence for evolutionary change.

Darwin's classification of the crustacean Subclass Cirripedia, being the first based on the evolutionary principle of common descent, remains an authoritative text. Since Darwin's time, numerous new species have been described, and well over twelve hundred barnacle species are currently placed into three main groups: Thoracica, Acrothoracica, and Rhizocephala.

The Thoracica group contains the great majority of barnacles, and includes all of the forms studied by Darwin. These animals are always attached to a hard substrate, such as rocks, shells, floating wood, backs of turtles, sea snakes (see page 60), whales, and, of course, ship hulls. They affix by means of an extraordinarily strong adhesive secreted by cement glands that form on the top of the head of the settling (cyprid) larva. In stalked or pedunculate barnacles, the cement glands attach to a long muscular stalk, while in the sessile or acorn barnacles, the cement glands are incorporated into a flat calcified plate. A ring of additional plates surrounds the body and is homologous with the carapace of other crustaceans. Inside the modified carapace, the animal lies on its back with its feeding limbs (cirri) projecting upward. Once attached, adult barnacles remain in place for life, filter-feeding by extending feathery cirri into the water column and drawing plankton and detritus inward for consumption.

Acrothoracicans are extremely small burrowing forms that lack external plates and spend their entire adult lives embedded in the shells of mollusks, corals, and sea stars. And Rhizocephalan barnacles are parasites of decapod crustaceans (see page 104) and so anatomically reduced that their relationship to other barnacles is only recognized through examination of their larvae.

3.

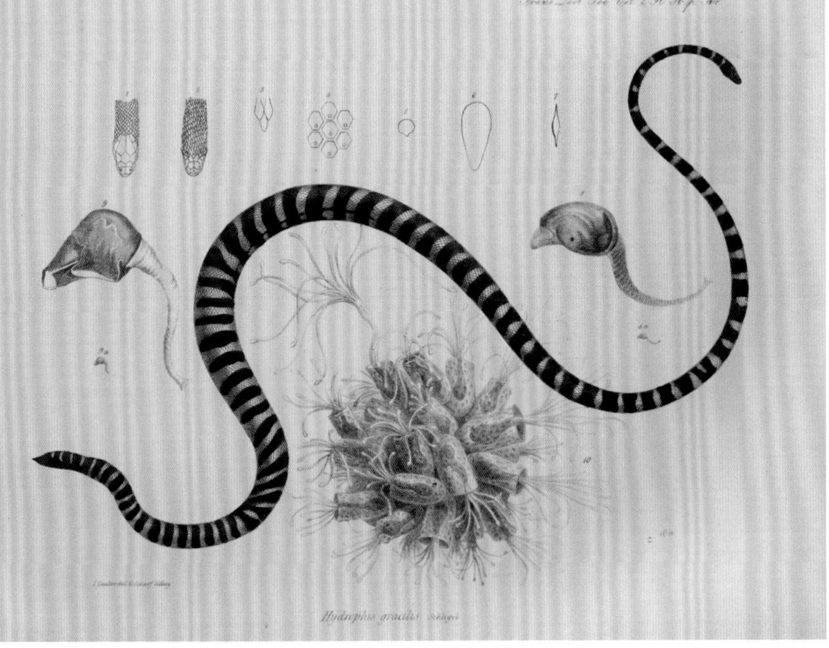

4.

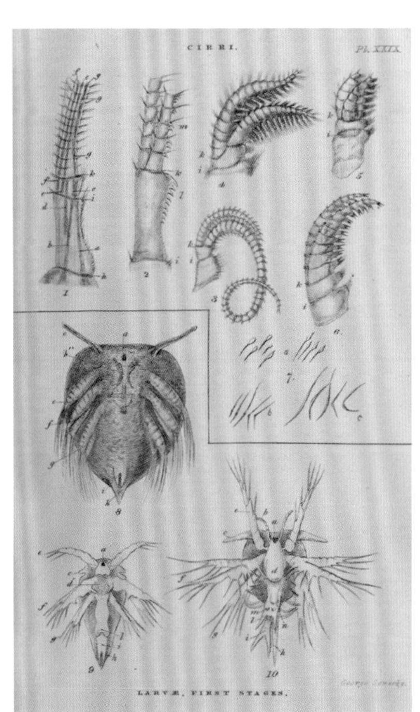

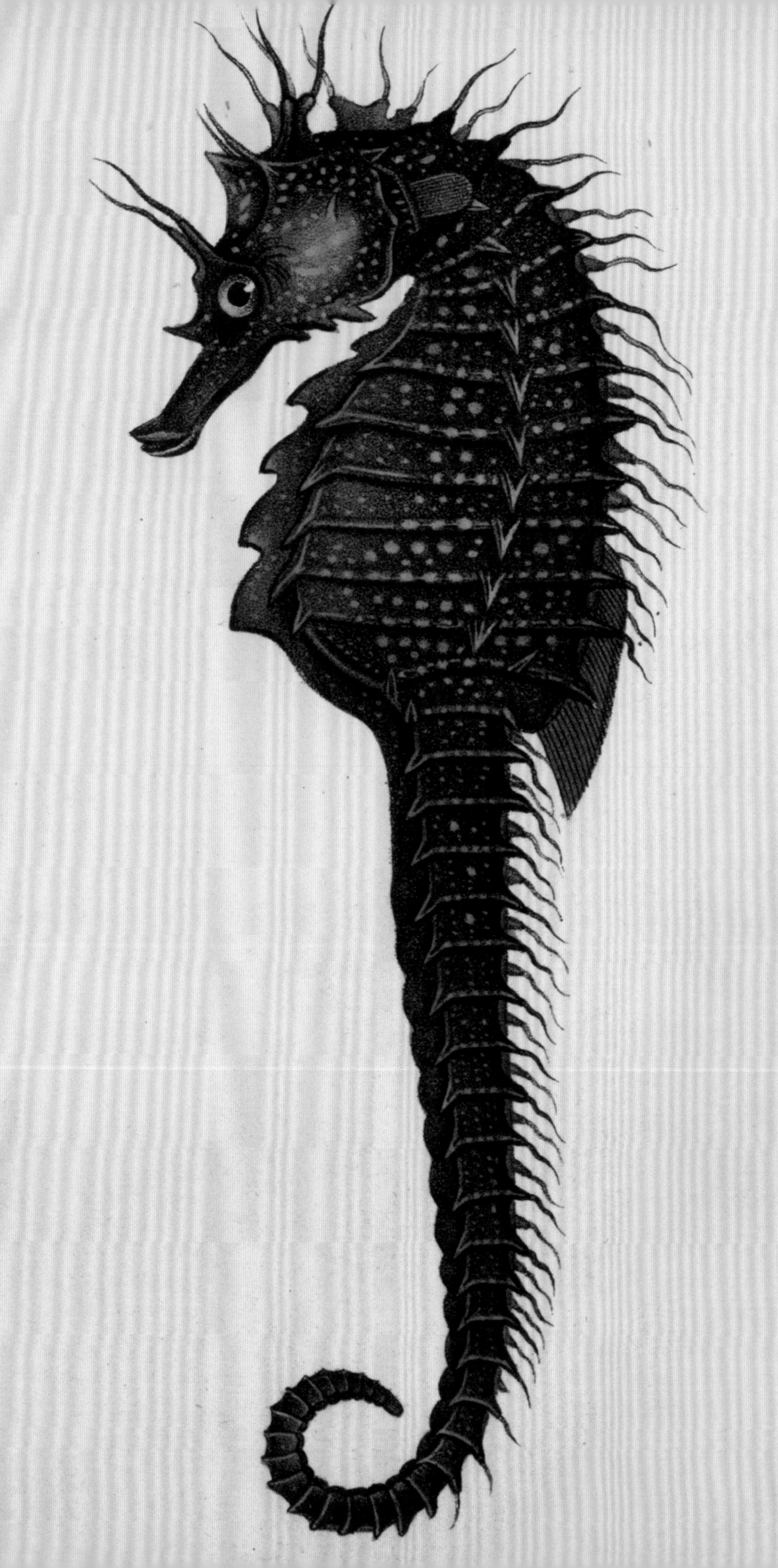

Schinz's Seahorse

Author
Heinrich Rudolf Schinz (1777–1861)

Title
Naturgeschichte und Abbildungen der Fische: nach den neuesten Systemen zum gemeinnützigen Gebrauche entworfen und mit Berücksichtigung für den Unterricht der Jugend

(Natural history and illustrations of fishes: following the newest systems developed for the general purpose of—and with special emphasis on—the instruction of youth)

Imprint
Schaffhausen: F. Brodtmann, 1856

Figure 4
Author: Albert C. L. G. Günther (1830–1914)
Title: *On the pipe-fishes belonging to the genus Phyllopteryx* (Proc. Zool. Soc. London)
Imprint: London: The Society, 1865

1. Seahorses belong to the genus *Hippocampus*, a name derived from the Greek *hippos* meaning horse and *kampos* for sea monster. Far from monsters, these charming fishes are today greatly threatened by overharvesting for traditional Asian medicine, curios, and the pet trade.

Heinrich Rudolf Schinz was a pioneer of the study of Swiss natural history and a prolific popularizer of science who published many beautifully illustrated compendia of animal life that reached a wide audience across the continent. He was born the son of a pastor in Zurich, and acquired a passion for nature from his father, a keen observer and collector of natural history treasures. Schinz was orphaned in early childhood and raised by wealthy relatives, and after attending school in Zurich, he left to study medicine. In 1798, at the age of just twenty-one, he completed his studies in Jena and traveled to Paris. A few years later he returned to Zurich, where he established a medical practice.

By 1804 he was teaching physiology and natural history at the Zurich Institute of Medicine, and in 1815 he cofounded the Swiss Naturalist's Society, where he played an active role in building interest in the local flora and fauna. Schinz's attentions turned increasingly to natural history, publishing important studies of the mammals and birds of Switzerland, and assuming responsibility for the zoological collections of the Naturforschende Gesellschaft in Zurich. Schinz was unstinting in his efforts for the society, actively building and curating its collections, and fund-raising for the society and for permanent storage facilities for its rapidly growing collections.

In 1833 the University of Zurich was founded and Schinz was appointed professor of zoology. He also played an increasingly active role in civic affairs, particularly those related to the education of the town's young people.

While highly regarded in his native Switzerland, Schinz also attained international recognition through the many beautifully illustrated, large-format compendia of animal life that he published from early 1820 through the late 1850s. These extremely popular publications—focused mainly on vertebrate diversity, particularly of mammals and birds—were based mainly on the works of others, which Schinz compiled and edited into accessible and attractive volumes. In collaboration with the natural history artist and engraver Karl Joseph Brodtman (1782–1862), many original illustrations were redrawn and reorganized by Schinz to reflect the prevailing classifications of the day.

Naturgeschichte und Abbildungen der Fische is one of these compilations, and in this case Schinz followed a Linnean classification for fishes. The work borrowed heavily from authorities such as Marcus Elieser Bloch (1723–1799), Baron Cuvier, Achille Valenciennes (1794–1865), and others from whom he gathered illustrations and compiled descriptive text for his selection of the fishes. Schinz's attractive two-volume publication is accompanied by ninety-seven hand-colored copperplate engravings, featuring more than 320 beautifully rendered fishes. It includes descriptive text and literature citation for each species, identified by its scientific epithet as well as by German and French vernacular names.

Schinz's beautiful image of Lacepede's crested oarfish (*Lophotus lacepede*) that accompanies this essay, for example, is based on a drawing in Baron Cuvier and Valenciennes's magnificent *Histoire naturelle des poissons*. Unfortunately, the origin of his leafy seahorse, identified as *Hippocampus foliatus*, is less clear, and Schinz uncharacteristically does not provide an authority for the image.

Schinz was already seventy-nine years old by the time his *Naturgeschichte und Abbildungen der Fische* was published. Although he appeared to have recovered well from a severe stroke he had had some six years earlier, he began to suffer from a series of small strokes that left him paralyzed and almost unable to speak. Despite his dismal physical condition, he maintained full mental clarity and an "indestructible gaiety" until his death in March 1861 at the age of eighty-four.

Seahorses and the closely related pipefishes, along with the leafy and weedy seadragons, are grouped together in the marine fish Family Syngnathidae, named for the tiny fused jaws that are located on the end of their elongate, tubular snouts. About fifty-four species of seahorses in the genus *Hippocampus* live in shallow sheltered waters mainly around sea grasses, coral reefs, and mangroves. They are highly modified fishes that have no scales and instead are encased in rigid bony plates that ring their bodies. As a result, they are poor swimmers, propelled only by a dorsal fin and a tiny pair of pectoral fins on each side of the head. Remarkable for fishes, they have no caudal fin and their tails are fully prehensile, allowing them to grasp onto seaweed fronds or coral heads, where they spend most of their time beautifully camouflaged, awaiting the approach of small crustacean prey. Once a prey floats into range, the seahorse, with its highly flexible neck, can precisely position its elongate tubular snout and rapidly suck the animal into its mouth.

All syngnathids share an unusual reproductive biology: it is the males that brood the eggs laid by the females. The male seahorse has a special pouch on its belly, and the female lays her eggs into the pouch. After fertilization, the eggs hatch and develop, well protected inside this pouch. After anywhere from ten to forty-five days, depending on the species, the male releases (gives birth to) hundreds of tiny, fully formed seahorses into the surrounding water. Once the babies are released, they are able to fend for themselves and the male provides no further care.

The seahorse that accompanies this essay was identified by Schinz as the leafy seahorse (*Hippocampus foliatus*), a name today considered to be a synonym (see page 68) for the weedy seadragon (*Phyllopteryx taeniolatus*). However, the fish that Schinz illustrated is clearly not a weedy seadragon, and instead looks rather like *Hippocampus breviceps*, the short-head or knobby seahorse. Both species are from Australian waters and it seems that Schinz, who probably never saw either, may have confused the two.

2. Many of Schinz's illustrations, such as this charming rendering of a mesopelagic crested oarfish (*Lophotus lacepede*), were taken from the works of others and redrawn and recolored by Brodtman.

3. This plate of a weedy seadragon (*Phyllopteryx taeniolatus*), published in 1865 by Albert Günther, was based on a drawing by Ferdinand Bauer, who accompanied Captain Flinders on his voyage to southern Australia.

Der Lacepedische Buschfisch. *Lophotes cepedianus.* *Lophote cépédien.*

2.

P. Z. S. 1865. Pl. XIV.

Phyllopteryx foliatus.

G.H. Ford. W. West imp.

3.

Anemones and Aquariums

Author
Philip Henry Gosse
(1810–1888)

Title
Actinologia britannica: a history of the British sea-anemones and corals

Imprint
London: Van Voorst, 1860

Figures 2 & 3
Author: Philip Henry Gosse
(1810–1888)
Title: *The aquarium: an unveiling of the wonders of the deep sea*
Imprint: London: J. Van Voorst, 1854

1. The acclaimed artist and lithographer William Dickes developed a new process for color lithography using copper plates, which produced the stunning naturalistic compositions that accompany Gosse's treatise.

Philip Henry Gosse was a brilliant, if somewhat eccentric, nineteenth-century English naturalist, writer, and theologian. Throughout his career, he published on a wide range of subjects including ornithology, herpetology, entomology, and even egyptology. After undergoing a religious conversion while traveling in Canada and the southern United States in the early 1830s, he produced numerous religious texts, and his religious views were expounded upon in many of his later natural history works. Despite wide-ranging interests and achievements, Gosse was perhaps best regarded for his contributions to marine biology, most particularly for his groundbreaking study of British sea anemones and corals published in 1860.

This beautifully illustrated work provided a refreshing new take on the monographic study of living organisms. It was published toward the end of his career, and Gosse resolved to produce a comprehensive study that was not only scientifically accurate and precise, but also one that "a student can work with." He was convinced that only by studying living animals in nature could their characteristics and discriminating features be fully understood and represented to the public. Highly critical of the work of others, he believed he had "received little aid—I may say almost literally none—from my predecessors." Of the classic *Histoire naturelle des coralliaires* by his French contemporary, Henri Milne-Edwards (1800–1885), he wrote that while it was a "work of immense research, labour, and patience, it bears evidence in every page of being the product of the museum and the closet, not of the aquarium and the shore," and that of many of the animals described "the learned author evidently had no acquaintance—or next to none—and hence he has merely reproduced the words of his authorities in all their vagueness, while the distribution of species into genera and families appears so full of manifest error to one personally familiar with the animals in a living state, that I have not attempted to follow his arrangement."

Instead, Gosse turned to nature. For eight years, he scoured the British coastlines making meticulous observations of the numerous marine communities that inhabited the island's rocky seashores. Many of the specimens that Gosse studied were taken to his home in St. Marychurch, Devon, and placed into aquariums that he had had constructed, and where he could continue to make observations and sketches of the living animals. Interestingly, Gosse apparently so disdained the official Church of England that he even rejected the use of the prefix "Saint" in St. Marychurch, instead claiming residence in the nearby town of Torquay. Nonetheless, despite some unorthodox religious views, Gosse was not only an excellent observer and recorder of nature, he was also a highly skilled draftsman. It was based on Gosse's own original drawings, sketches, and compositions that the highly

2.

acclaimed engraver William Dickes (1815–1892) produced the extraordinarily beautiful color plates that illustrate this volume.

A few years prior to the publication of *Actinologia Britannica*, Gosse had coined the term "aquarium," and produced a highly popular handbook with practical instructions for constructing, stocking, and maintaining a marine tank in the home. This publication was enthusiastically received, spurring much public interest in marine life, and is credited with setting off the Victorian craze for keeping aquaria and collecting living organisms at the seashore.

Anemones are closely related animals that, together with true jellyfish (see page 140), box jellies (see page 49), and hydrozoans (see page 44), form an exclusively aquatic phylum called Cnidaria. Anemones are predators that feed mostly on small fishes and crustaceans. These they capture using numerous tentacles armed with venomous stinging cells called nematocysts, which surround a single, upward-facing opening on the oral disc. Unlike most other cnidarians, anemones lack a free-swimming medusa stage and spend their entire life cycle attached to the substrate.

Although anemones are capable of some movement and can release themselves from the substrate and move short distances, they remain in one place for most of their lives. Individual anemones can be very long lived, commonly reaching sixty to eighty years of age or even older if they don't succumb to their main predators, the sea slugs (see page 147) and sea stars (see page 64).

2. Gosse's aquariums, like this grand "fountain aquarium" illustrated in his popular book, *The aquarium*, have been credited for sparking the Victorian craze for rock pooling and collecting seashore creatures.

3. The ancient wrasse (*Labrus bergylta*) is featured on the frontispiece of Gosse's *The aquarium*. He considered it one of the hardiest fish to keep and represented it lurking under a projecting ledge of rock, a situation, he says, it loves to haunt.

4. The so-called parasitic anemones (*Calliactis parasitica*) featured in this plate are often found attached to hermit crab shells. Their relationship is not parasitic, but is mutualistic, as the stinging anemone protects the crab while, in turn, eating scraps of the hermit's food.

5. This beautiful composition features an assortment of anemones alongside a large colony of the coldwater stony coral (*Lophelia pertusa*). Stony corals and anemones are closely related members of the cnidarian Subclass Hexacorallia.

THE ANCIENT WRASSE

3.

1.8. SAGARTIA NIVEA. 5. S. TROGLODYTES. 7. S. ICTHYSTOMA.
2.3.4. S. MINIATA 6. S. PARASITICA 9.10. S. ORNATA

4.

1. LOPHELIA PROLIFERA. 5. ZOANTHUS COUCHII. 9. PHYLLANGIA AMERICANA
2. PEACHIA TRIPHYLLA. 6. PARACYATHUS TAXILIANUS. 10.11. BALANOPHYLLIA REGIA
3. SPHENOTROCHUS WRIGHTII. 7. P. PTEROPUS. 12.13. CYATHINA SMITHII
4. S. MAC ANDREWANUS. 8. P. THULENSIS.

5.

Corail épanoui
de grandeur naturelle et grossi.

H.L.D. ad nat. del. *Librairie J.B. Baillière et Fils, Paris* *Annedouche sculp.*
Imp. A. Salmon, R. Vieille Estrapade, 15.

Precious Red Coral

Author
Henri de Lacaze-Duthiers
(1821–1901)

Title
Histoire naturelle du corail: organisation, reproduction, pêche en Algérie, industrie et commerce

(Natural history of coral: organization, reproduction, fishery in Algeria, industry and trade)

Imprint
Paris: J. B. Baillière, 1864

1. This beautiful branch of red coral (*Corallium rubrum*) was described by Lacaze-Duthiers as being in full bloom (*épanoui*) with each coral animal (polyp) fully extending the eight tentacles surrounding its mouth, as if feeding.

The career of Félix Joseph Henri de Lacaze-Duthiers spanned a time of much discovery in the rapidly advancing field of marine biology. His influence was profound, not only for numerous scientific contributions, but also for the influential role he played as a revered teacher and as the driving force behind the establishment of two of his nation's most important marine biology laboratories.

Lacaze-Duthiers studied and practiced medicine in Paris, but soon came under the thrall of zoology. After a trip exploring marine life in the Mediterranean Balearic Islands, he returned to Paris to work as assistant to the eminent biologist Henri Milne-Edwards (1800–1885), who was chair of entomology at the Paris Natural History Museum. By 1854 Lacaze-Duthiers was appointed professor of zoology at the University of Lille and produced a series of well-received studies of marine invertebrates. A consummate anatomist known for his meticulous dissecting skills, he rather famously refused to employ modern anatomical techniques—such as the newly developed method of serial sectioning specimens for microscopic analysis. He nonetheless was a forceful advocate for a modern experimental approach to zoology.

Unlike many of his contemporaries, Lacaze-Duthiers appreciated the importance of studying live animals, and regularly left Paris for the Mediterranean coast and islands. In 1858, while on Minorca, he noticed a fisherman marking his clothes with the "purple of the ancients." Upon further investigation, he observed that the glandular secretions of certain marine snails, when exposed to sunlight, became a deep purple. One year later, he published *Mémoire sur la pourpre,* establishing that the purple dye so prized by the ancient Romans was derived from the glands of these Mediterranean snails.

By this time, the French occupation of Algeria had been ongoing for thirty years, and Lacaze-Duthiers was invited to study the corals of the colony, particularly the highly valued red coral (*Corallium rubrum*) with a view toward establishing a well-regulated French trade in this precious commodity. The task had first been offered to Jean Louis de Quatrefages (1810–1892) of the Paris Natural History Museum, but Quatrefages had demurred, suggesting that Lacaze-Duthiers would make an ideal substitute. And so on October 1, 1860, Lacaze-Duthiers set sail for Algeria, where he was to stay for a year observing, collecting, and carefully rearing corals.

He then returned to France, although with typical punctiliousness, he revisited Algeria twice more to confirm many of his initial observations and, over a two-year period, wrote what was to become one of his most celebrated works, the classic *Histoire naturelle du corail.* Lacaze-Duthiers provided a scholarly history of the study of corals, with an account of their biology and habitats. He made many novel observations on coral physiology, reproduction, and larval metamorphosis, and provided a review of the red coral trade, calling for legislative reform to regulate it.

The memoire is beautifully illustrated with twenty multifigure plates, each figure drawn with great care by the author himself. It was dedicated to his mother, who must have died while he was in Algeria, as he ends with the poignant exclamation, "Ah! If I have not been able to receive your last embraces, my dear revered mother, then I can at least offer this dedication to your noble and generous soul, with love and regret!"

After publication of the *Histoire*, Lacaze-Duthiers was appointed professor of zoology at the Paris Natural History Museum. In 1868 he moved to the Sorbonne as chair of zoology and by 1871 was elected to the Paris Academy of Sciences, later becoming its president. As a teacher, Lacaze-Duthiers was keenly aware of the need for French students to study living marine organisms and, using his powerful academic position, argued for the establishment of a marine station at Roscoff on the Brittany coast. In 1872 the Roscoff station opened, initially as an annex of the Sorbonne, but Lacaze-Duthiers welcomed students and international researchers alike. In 1879 he began pressing for a second station, this time on the Mediterranean coast, where very different marine environments could be studied, and in 1881, France's second marine research station, the Laboratoire Arago in Banyuls-sur-Mer, was established. Preferring the weather and his beloved Mediterranean, Lacaze-Duthiers continued working in his laboratory at Banyuls until a few days before his death in July 1901.

Corals are members of the Phylum Cnidaria and are closely related to sea anemones (see page 92). The red, or precious, corals are octocorals and each of the hundreds of tiny polyps (coral animals) that make up a colony has a mouth surrounded by eight tentacles; the polyps of reef-building, stony corals (see page 52) have six tentacles. Red corals belong to the soft coral Order Alcyonacea. However, the famous Mediterranean red coral (*Corallium rubrum*) and other precious *Corallium* species differ from other soft corals (see page 22), which secrete a flexible, protein-based skeleton, in producing a very hard, fully calcified exoskeleton. It is this extremely durable and beautifully colored outer skeleton that has made them such a prized commodity throughout human history. Red coral jewelry has been found in prehistoric burial sites from around Europe, and already by the first millennium a thriving trade between the Mediterranean and India was in existence.

2. Lacaze-Duthiers made seminal contributions to the understanding of red coral reproduction. Here, he illustrates male coral polyps releasing clouds of sperm that are negatively buoyant and sink into the water of the container.

3. The heavily calcified skeleton of red coral is itself colored in various hues of red by carotenoid pigments. It is this durable red skeleton, which can be polished to a glossy sheen, that makes it such an attractive material for decorative use.

4. Lacaze-Duthiers discovered that *Corallium rubrum* sperm fertilizes eggs inside the female polyps. Once developed, their larvae are released into the water column as illustrated in this plate.

LACAZE-DUTHIERS
Le Corail.
Pl. XII.
Mâles lançant la semence.
Lait ___ Capsule de l'œuf.
Librairie J.B. Baillière et Fils Paris.

2.

LACAZE-DUTHIERS.
Le Corail.
Pl. XX.
Variétés du Corail.

3.

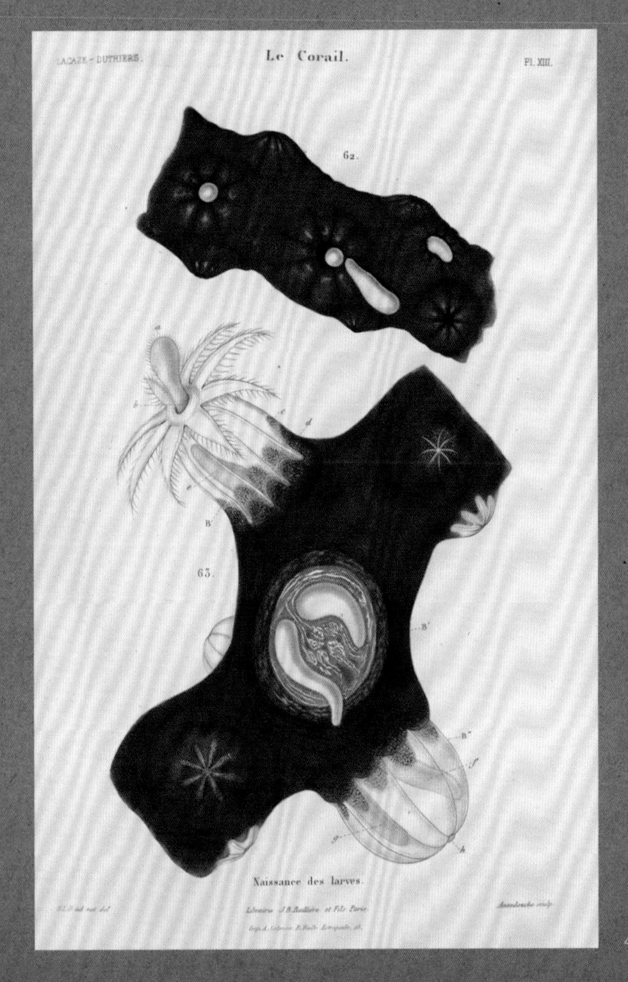
LACAZE-DUTHIERS.
Le Corail.
Pl. XIII.
Naissance des larves.
Librairie J.B. Baillière et Fils Paris.

4.

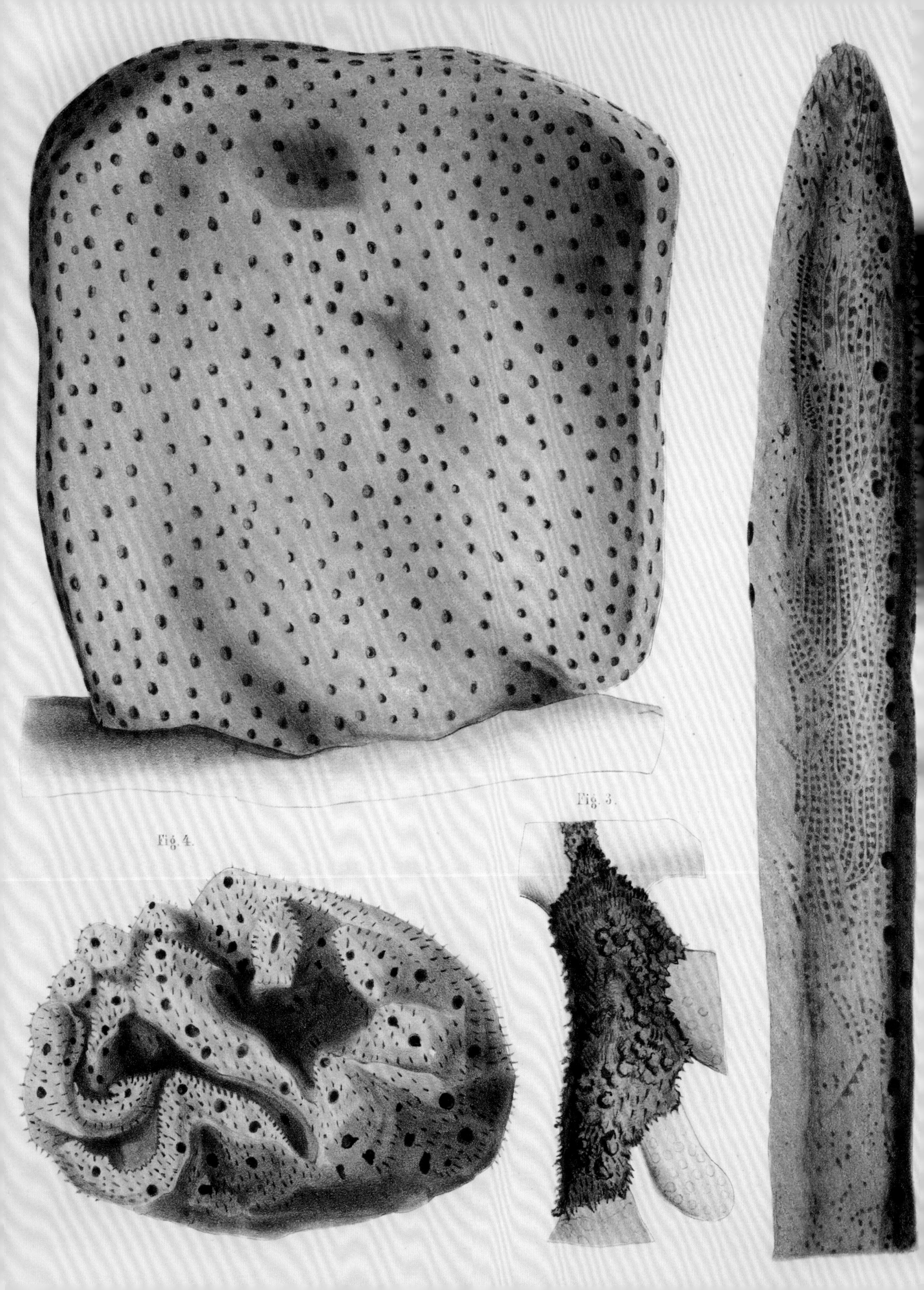
Fig. 3.
Fig. 4.

The Sponges of the Caribbean

Authors
Placide Duchassaing de Fontbressin (1819–1873)
Giovanni Michelotti (1812–1898)

Title
Spongiaires de la mer caraïbe

(Sponges of the Caribbean Sea)

Imprint
Haarlem: Société hollandaise des sciences, 1864

1. Sponges are an ecologically important and highly diverse group of marine animals. In this work Duchassaing and Michelotti described numerous new species from the shallow waters of the Caribbean.

Édouard Placide Duchassaing de Fontbressin was born to an Antillian Creole family of planters on the French island of Guadeloupe. As a young man, his parents dispatched him to Paris, and there he studied medicine, zoology, and geology. After graduation, Duchassaing returned to Guadeloupe and established a medical practice, but spent much of his time researching the island's flora and fauna. He soon extended his studies with travels throughout the neighboring islands and to Panama. In Panama he worked particularly on the flora of the isthmus and began a correspondence with the German botanist Wilhelm Gerhard Walpers (1816–1853), to whom he sent many specimens.

After several years on Guadeloupe, Duchassaing, by then familiar with all the islands in the region, decided to relocate to the more cosmopolitan commercial center of St. Thomas, which, although a former colony of the British, was at that time part of the Danish West Indies. In order to establish a medical practice on St. Thomas, Duchassaing was required to gain Danish credentials, so in the early 1850s he traveled to Copenhagen and obtained accreditation there.

In 1852 he returned to St. Thomas, where he was to be based for the next fifteen years, and began a series of studies focused on the in-shore marine invertebrates, particularly the corals and sponges of the region. Although working more or less in isolation, Duchassaing had established correspondence with a number of European naturalists and, in the winter of 1854–1855, he met with his longtime collaborator, the Italian geologist and paleontologist Giovanni Michelotti (1812–1898). Michelotti had a particular interest in fossil and living corals, and the two spent three months in the Antilles making observations of in-shore communities of living animals, searching for fossils, and assembling large collections of specimens, most of which would return with Michelotti to the Turin Museum of Zoology.

In the following years, Duchassaing and Michelotti published a series of papers on the corals and sponges of the Caribbean, together describing numerous species new to science. The first in the series was a memoire on the corals of the region, published in 1860 by the Royal Academy of Sciences in Turin. Four years later, their treatise on sponges, *Spongiaires de la mer caraïbe*, was published. In the preface to that work, they noted that although "sponges do not have freedom of movement, nor the strength or beauty of form and color observed in many other zoological classes, and they may appear, so to speak, as neglected bodies at the bottom of the water," they nonetheless "play a rather important role in the life of the marine fauna." They also pointed to the paleontological utility of microscopic examination of sponge "debris" (presumably fossilized remains) as an aid in understanding past geological epochs, a particular interest of Michelotti.

Duchassaing and Michelotti acknowledged that their study was by no means a complete inventory of Caribbean sponges but that they had increased the number of species known from the region by fivefold. And, unusual for biologists at this time, they stressed the importance of long-term studies of living sponges over different seasons and in the numerous habitats where they were found. Clearly they were well aware of the variability of form that individuals of the same species can assume when exposed to differing conditions of location, depth, and water current. Their treatise is accompanied by twenty-five hand-colored lithographic plates illustrating large numbers of sponges with some detail of their microscopic anatomy. The authors stressed that their drawings were made from specimens "in their natural state, that is to say while alive," because when removed from water, these animals not only changed their colors, but, regardless of how carefully dried, dramatically changed in form.

In 1867 Duchassaing left the Caribbean for the last time and settled in Périgueux in southwestern France, where he died six years later at the age of fifty-four. His longtime colleague and collaborator, Giovanni Michelotti, continued geological and paleontological studies in Italy, and retired in 1879. In 1880 he donated his important fossil collections to the Institute of Geology and Paleontology at the University of Rome, and relocated to Sanremo on the Mediterranean coast, where he died in 1898.

Sponges, although superficially plantlike, are primitive multicellular animals (metazoans) belonging to the Phyllum Porifera. This name refers to the numerous pores and channels that allow water to be pumped and filtered by specialized cells (choanocytes) through two loosely aggregated layers of body cells. Sponges are highly variable in size, shape, and color, and are found exclusively in aquatic environments. Of roughly fifteen thousand species, most are marine and are found from shallow waters to the deep ocean abyss, although they are particularly common in coral reefs and in-shore mangroves and sea grass beds.

Sponges are grouped into three separate classes depending on the composition of the elements (spicules) that form their supporting "skeleton": glass sponges (Class Hexactinellida) with siliceous spicules, calcareous sponges (Class Calcarea) with calcium carbonate spicules, and demosponges (Class Demospongiae) with protein fiber spicules. More than half of all sponges are demosponges, and this very diverse group includes some of the largest barrel sponges, which can attain girths of over 1 meter (c. 3 feet) in diameter. Other demosponges are the encrusting sponges, which spread over rocky surfaces and reefs, and the commercially important "bath" sponges.

Sponges play an important role in many marine ecosystems, where they filter and clean large volumes of water as well as recycle nutrients. They also contribute significantly to reef building, and their hollow bodies provide habitat and shelter for numerous other marine organisms.

Most of the Caribbean sponges described by Duchassaing and Michelotti belong to the large poriferan Class Demospongiae, a group that includes the familiar bath sponges (*Spongia*). The four plates selected here depict a range of demosponge genera illustrating the remarkable diversity of these animals:

2. *Agelas* and *Amphimedon.*

3. *Thalysias* and *Pandaros.*

4. *Pandaros* and *Spongia.*

5. *Terpios* and *Geodia.*

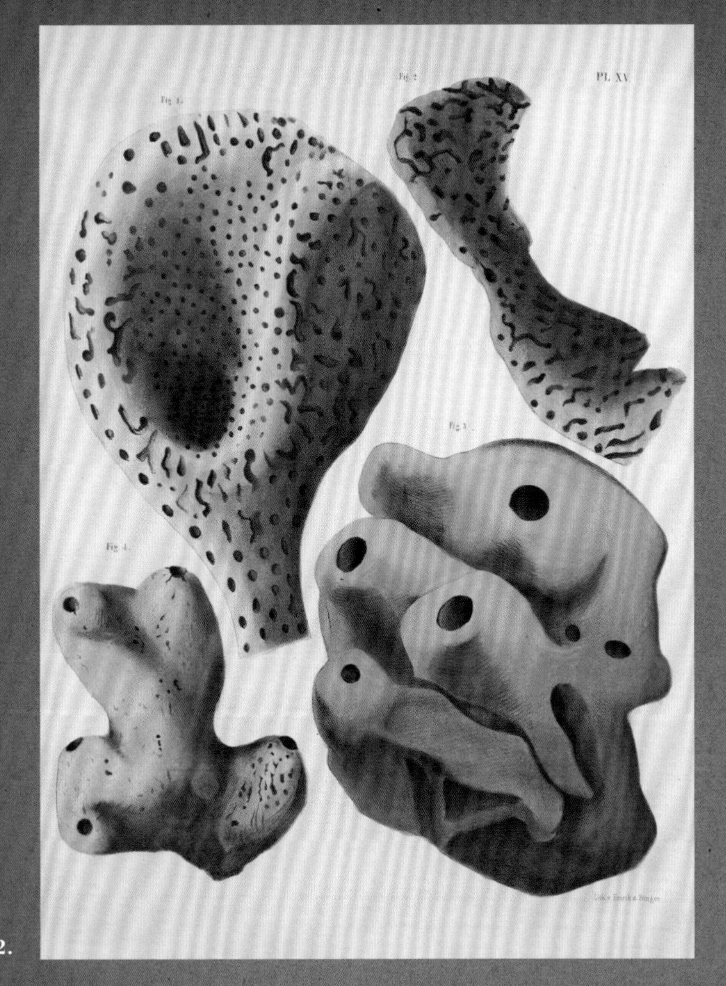

2.

3.

4.

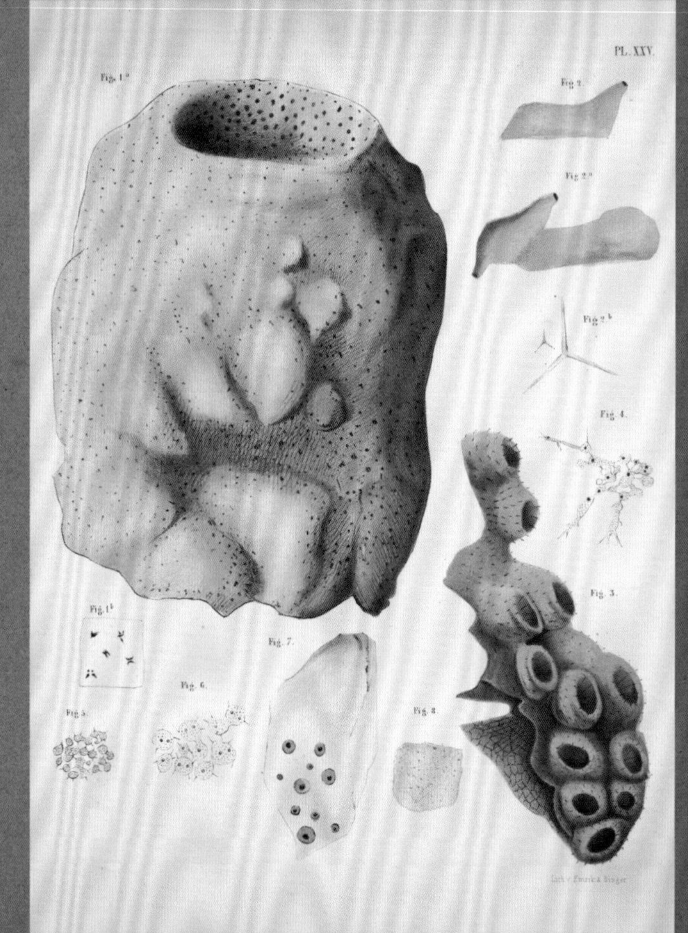

5.

D'Orbigny's Decapods

Author
Charles d'Orbigny
(1806–1876)

Title
Dictionnaire universel d'histoire naturelle: servant de complément aux oeuvres de Buffon, de G. Cuvier, aux encylopédies, aux anciens dictionnaires scientifiques, et résumant les traités spéciaux sur les diverses branches des sciences naturelles, etc.

(Universal dictionary of natural history: serving as a complement to the works of Buffon, G. Cuvier, and to encyclopedias, to old scientific dictionaries by summarizing the special treatises on the various branches of natural science.)

Imprint
Paris: Au bureau principal de l'éditeur, 1867–1869

1. This magnificent plate depicts a specimen of European lobster (*Homarus gammarus*) and is typical of the very high quality of illustrations accompanying D'Orbigny's monumental *Dictionnaire.*

During the late eighteenth and early nineteenth centuries, French naturalists published a number of influential "dictionnaires" of natural history. These impressive compilations were more akin to technical encyclopedias than to actual dictionaries, and provided large amounts of information about a wide range of topics and terms while summarizing the systematic knowledge of organisms of the time. One of the better known of these, and certainly the most beautifully illustrated, was the thirteen-volume *Dictionnaire universel d'histoire naturelle*, a monumental multiauthored compilation edited by Charles Henry Dessalines d'Orbigny, with whose name it is so intimately associated.

D'Orbigny was born in the small town of Couëron on the Loire, some 10 miles west of Nantes. His father was a town doctor, and a close friend of John James Audubon (1785–1851), who had also been raised in Couëron. Many years later, Audubon was reintroduced to Charles, whom he recalled having "held in my arms many times," and greeted him as his godson. D'Orbigny's elder brother, Alcide Charles Victor Dessalines d'Orbigny (see page 63), was to become a highly celebrated naturalist and pioneer explorer in South America. And, although perhaps overshadowed by his brother's greater fame and academic achievement, d'Orbigny nonetheless pursued a career in natural history. He had studied medicine in Paris, but had been more interested in geology and botany than in having a medical career and in 1834 attained a junior position in the geology department at the Paris Museum of Natural History, where his brother would ultimately have a chair of paleontology created in his honor. By 1837 he was appointed assistant naturalist at the museum, and he worked on many of the botanical specimens collected by his brother in South America as well as on some geological studies.

In 1839 the Paris publishing house of C. Renard circulated a prospectus for a *Dictionnaire universel d'histoire naturelle*, planned to comprise six to eight octavo volumes issued in 120 installments and edited by d'Orbigny. Without doubt, the choice of the thirty-three-year-old d'Orbigny, an assistant naturalist without academic standing, was a remarkable testament to his breadth of knowledge and scientific acumen. More than fifty of France's most celebrated savants contributed to the work, many of whom were d'Orbigny's superiors at the Paris Museum of Natural History including such notables as Isidore Geoffroy Saint-Hilaire (1805–1861), Antoine Laurent de Jussieu (1748–1836), Henri Milne-Edwards (1800–1885), and Achille Valenciennes (1794–1865). D'Orbigny himself provided an introductory discourse on the history of the natural sciences.

The *Dictionnaire* was completed ten years later, a monumental accomplishment ultimately consisting of thirteen volumes published in 150 installments, with a second edition "considerably augmented and enriched" appearing between 1867 and 1869. That edition was accompanied by "a magnificent Atlas of more than 300 engraved and colored plates" and it is these spectacularly prepared and executed plates that make the work such an important contribution, not just a testament to the science of the day, but also a remarkably successful means of popularizing that science.

As with the dictionary itself, the artwork was the product of many collaborators, among them some of the finest natural history artists and engravers of the day. These included artists such as Jean-Gabriel Prêtre (1800–1840), Paul Louis Oudart (1796–1850), and Louis Edouard Maubert (1806–1879). The naturalist Émile Blanchard (1819–1900), who in 1862 was to became chair of Crustacea, Arachnida, and Insects at the Paris Museum of Natural History, produced the extraordinary depictions of decapod crustaceans that accompany this essay.

Decapod ("ten-footed") crustaceans (shrimps, crabs, crayfish, and lobsters) together with amphipods, isopods, and stomapods (mantis shrimps), belong to the Malacostraca, the largest crustacean class, which includes more than twenty-five thousand species. All share a basic body plan consisting of a head (five segments), thorax (eight segments), and abdomen (six segments), but each may be modified in different ways.

In many decapods the anterior thoracic segments fuse with the head to form a cephalothorax, and five pairs of walking legs (pereiopods) are present on the last five thoracic segments. The abdominal appendages are called pleopods and are used for swimming and for brooding eggs; the terminal appendages form a tail fan with a pair of lateral uropods and a median telson bearing the anus. The head has a pair of antennae and antennules, stalked eyes, and numerous feeding appendages.

Based on features of their gills and legs, as well as on larval development, decapods are divided into two main Suborders: Dendrobranchiata (the prawns) and Pleocyemata (the lobsters, crayfish, true shrimps, and crabs). Among pleocyemates, lobsters are grouped with crayfish and their allies in the Infraorder Astacidea; these are recognized by claws (chelae) on the first three pairs of walking legs, with the first pair greatly enlarged. Clawed, or true, lobsters belong to the Family Nephropidae, and although the term lobster is also applied to the spiny, slipper, and squat lobsters, these are not astacids and are not closely related to true lobsters.

About fifty species are known worldwide, where they live mainly in in-shore waters, searching for food on the seafloor and sheltering among rocks and in burrows. Much of their activity takes place in murky water or at night and lobster vision is not acute; they rely on very long antennae for odor detection to locate prey. Most species are omnivores, feeding on a variety of invertebrates and algae, although many will also take carrion. In northern waters, two economically important species are the closely related Maine (*Homarus americanus*) and European (*Homarus gammarus*) lobsters. Both are slow growing and long-lived, with the Maine lobster attaining lengths of over 0.6 meters (c. 2 feet) and weighing more than 18 kilograms (c. 40 pounds) over a life span of more than fifty years.

Blanchard's illustration is both beautiful and highly accurate, and the specimen he drew readily identifiable as a mature European lobster. But sex determination is not possible from this illustration as the characteristic features of each gender are only visible on the ventral surface. In males, the first abdominal pleopods are modified for copulation and are rigid, while in females they are soft and feathery.

2. The second pair of periopods (walking legs) of this beautiful shrimp is greatly enlarged, a characteristic of members of the fresh-water palaemonid genus *Macrobrachium*.

3. This magnificent mantis shrimp, *Lysiosquillina maculata*, attains lengths of up to 40 centimeters (c. 16 inches). Mantis shrimp belong to the Order Stomatopoda, sometimes called "thumb splitters," as many species can inflict considerable damage with rapid, powerful strikes or stabs of their enlarged front legs.

4. The blue swimmer crab (*Portunus pelagicus*), a highly prized food throughout the Indo-Pacific, is an excellent swimmer. Swimmer crabs (Family Portunidae) use a pair of flattened paddlelike periopods to rapidly scull through the water. Males of this species are bright blue, while females are green.

DICT. UNIV. D'HIST. NAT.

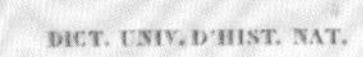
Crustacés. PL. 1.

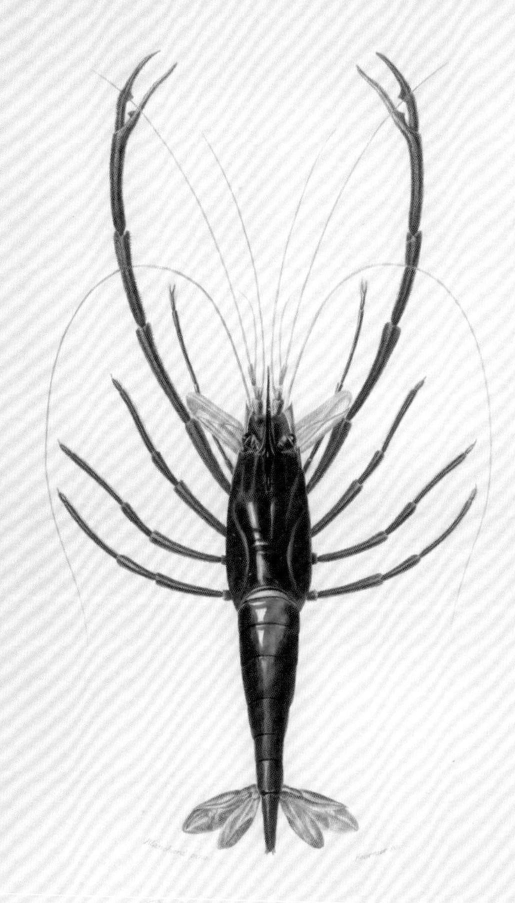

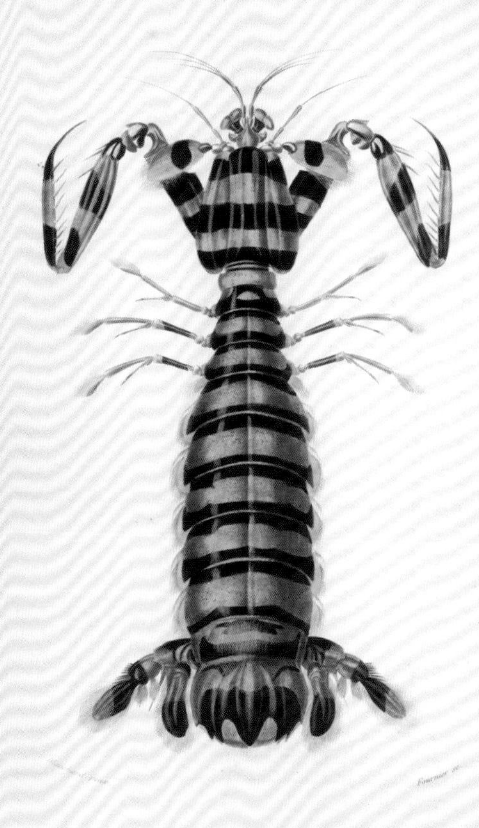
DICT. UNIV. D'HIST. NAT.
Crustacés. PL. 3.
Squille maculée. (Squilla maculata, Lamk.)

3.

Blanchard pinx.
Fournier sc.
Lupée pélagienne. (Lupa pelagica, Lin.)
Bulliau imp.

4.

The Mollusks of Japan

Author
Carl Emil Lischke
(1813–1886)

Title
Japanische Meeres-Conchylien: ein Beitrag zur Kenntniss der Mollusken Japan's, mit besonderer Rücksicht auf die geographische Verbreitung derselben

(Japanese marine shells: a contribution to the knowledge of the mollusks of Japan, with particular attention to their geographical distribution)

Imprint
Cassel: Theodor Fischer, 1869–1874

Carl Emil Lischke trained as a lawyer in Berlin, and upon return to his hometown of Stettin, then the capital of the Prussian Pomerania, he was appointed deputy town magistrate. In 1847 he left for the United States as attaché to the Prussian Embassy in Washington, D.C. Returning to Germany in 1850, he became mayor of the industrial city of Elberfeld, a position he held until 1873, when ill health required that he retire from public service.

Despite many civic duties and responsibilities, Lischke maintained a lifelong interest in the natural sciences, particularly the study of mollusks and their shells. He traveled widely in Europe and northern Africa, and in his later years visited Ceylon (Sri Lanka) in the East Indies. During these travels, he assembled large collections; many new mollusk species were described based on shells that he collected. Without doubt, Lischke's greatest scientific achievement was the publication, between 1869 and 1874, of the meticulously researched three-volume *Japanische Meeres-Conchylien.* In these volumes, Lischke provided detailed accounts, accompanied by many stunning plates, of the shells then known from waters of the Japanese archipelago. At the time, unlike in many other regions of the world, Japan's shells were very poorly known to European collectors. The reason for this, Lischke explains in the preface to the first volume, was the long cultural and political isolation of Japan during the Edo period (1608–1868).

During the Edo, only the Dutch—residing between 1641 and 1854 at Dejima, a trading post in the Nagasaki Bay—had contact with the Japanese and had any opportunity for scientific study in the region.

One notable collector during the "Dutch period" was the physician and traveler Philipp Franz von Siebold (1796–1866). Siebold spent six years in Japan, first at Dejima; later, he was granted limited access to the mainland, where he amassed large collections of plants and animals, including shells. Siebold's primary interests were botanical, and he famously smuggled seeds of the tea plant out of Japan, and established tea culture on the Dutch-held island of Java.

In 1853, after more than two hundred years, the Dutch trading monopoly in Japan was challenged by the arrival of Commodore Matthew C. Perry and his famous U.S. expeditionary fleet. By 1854 Perry and the ruling Tokugawa shogun had signed the Kanagawa Treaty, and Japan's isolation was over. A few years later, in the *Japanische Meeres-Conchylien*, Lischke gratefully acknowledged the pioneering efforts of Siebold and of the naturalists who accompanied Perry's expeditions and who in 1856 had published a *Report of the shells collected by the Japan Expedition, under the command of Commodore M. C. Perry, U.S.N., together with a list of Japan shells.* Lischke incorporated much of that material in his compendium.

1. The large predatory sea snail that made this beautiful shell was a Saul's Japanese triton (*Charonia lampas sauliae*). Tritons are active hunters, particularly of starfish, and have been observed actively chasing them.

Beyond being a careful cataloging and description of the shells of Japan, Lischke's study was also the first summary of their geographical distributions. He noted, for example, that a number of shells found in Japan were also present in the Mediterranean—an unexpected finding—and it was the zoogeographical aspects of his work that other conchologists deemed the most interesting. His friend, the eminent British naturalist and explorer J. Gwyn Jeffreys (1809–1885), reviewed Lischke's work in *Nature* magazine in 1870, remarking that it was unfortunate that too little was known "of the present course of those great currents which traverse the ocean in every direction, to be able to explain satisfactorily the geographical distributional of the marine fauna." He commended Lischke for compiling the distributional data that highlighted such patterns. Gwyn Jeffreys concluded his review with the interesting observation that "the author [Lischke] finds time not only for his onerous public duties, but also for good scientific work; so that in other countries besides our own, writers on natural history are not confined to the class of paid professors."

Although an amateur who lacked formal training, Lischke's place in the history of conchology was formally recognized when he was awarded an honorary doctorate by the University of Bonn in 1868.

Conchology is the study of the shells, rather than of the animals that made them, and is an early branch of malacology, the scientific study of mollusks. Because of their ubiquity, the edibility of many of their inhabitants, and their beauty, shells have long captured human interest, and shell necklaces and ornaments are common even in the earliest archeological sites from around the globe. Since the publication in 1685 of one of the first scientifically organized shell compendia, *Historia conchyliorum*, authored by Martin Lister (1639–1712), many hundreds of conchological treatises containing literally thousands of beautiful illustrations have been produced.

While today most malacological research is focused on the study of the living animals that make up the Phylum Mollusca, shell collecting remains a popular hobby. Around the world, numerous national conchological societies and local clubs exist and encourage the lawful, nondestructive collection and exchange of shells.

2. This plate depicts a variety of Japanese bivalves with a beautiful rendering of the shell valves of the mussel *Crenomytilus grayanus*, with its inner lining of iridescent nacre, or mother-of-pearl. The smooth layer of nacre helps protect the mussel's soft tissues against parasites and abrasion.

3. Triton shells (Family Ranellidae) have been valued since ancient times and are named after Triton, the messenger of the sea, who was the son of Poseidon. Triton is often depicted blowing a trumpet modeled from the shell of one of these large predatory snails.

4. A magnificent specimen of *Mytilus coruscus*, a hard-shelled edible mussel of the Family Mytilidae. Removed from the shell are the characteristic byssal threads that most mussel species use to firmly anchor themselves to the substrate.

5. Cultured by the Japanese for centuries, the Miyagi oyster (*Crassostrea gigas*), was introduced into the United States in the 1920s and is now the most widely farmed and commercially important species of oyster (Family Ostreidae) worldwide.

Taf. X.
1.
3.
4.
2.
5.
6.
5 a.
7.
8.
5 b.
1.2. Macha divaricata Lischke. - 3.4. Pecten yessoensis Jay, junior. - 5.6. Cecella chinensis Deshayes.
7.8. Mytilus Dunkeri Reeve.

2.

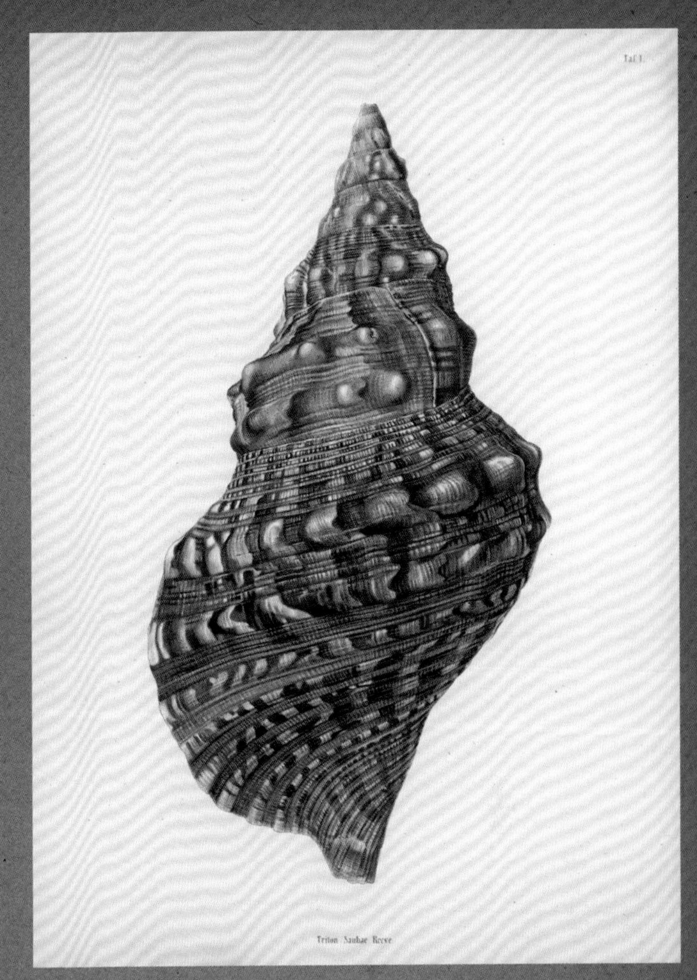
Taf. I.
Triton Sauliae Reeve

3.

Taf. XI.
1.
2.
1.2. Mytilus crassitesta Lischke

4.

Taf. XIV.
1.
2.
1.2. Ostrea gigas Thunberg, varietas

5.

McIntosh's Monograph of Marine Worms

Author
William Carmichael McIntosh (1838–1931)

Title
A monograph of the British marine annelids

Imprint
London: Ray Society, 1873–1923

The eminent Scottish physician and marine biologist William Carmichael McIntosh was the driving force behind the establishment of the first marine biology station in the British Isles. During the mid- to late-nineteenth century, a surge of interest in marine biology had seen the establishment of a growing number of permanent marine biology stations on the European continent (see pages 96 and 127). In 1882 McIntosh, then lecturing at the University of St. Andrews, bemoaned the fact that no such research facility existed in all of Britain. He argued that St. Andrews would make the perfect site for the first British marine station, with "its proximity to the sea and its quietude—so conducive to study—and the valuable library and museum of the University on the one hand and, on the other, the fine stretch of sand on which so many rare marine forms are thrown by storms, render it pre-eminent in this respect."

Just fourteen years after that address, with the financial support of the wealthy lawyer and naturalist Charles Henry Gatty, whom McIntosh had met at the Ray Society, the Gatty Marine Laboratory of the University of St. Andrews was established in 1896 with McIntosh as its first director. This was a position he was to occupy until 1917 when his successor, Sir D'Arcy Wentworth Thompson (1860–1948), took over as the chair of Natural History at St. Andrews.

It was many years earlier, however, while employed as a psychiatrist and superintendent of the Murthly mental hospital just outside of Perth, that McIntosh had begun his groundbreaking studies of marine worms—among the most common of all the denizens of in-shore coastal communities. The first volume, which described the strange nemertean, or ribbon, worms (see page 132), was published by the Ray Society in 1873, and in the following fifty years, three additional multipart volumes appeared, each describing and illustrating in exquisite detail the numerous families and species of polychaete, or bristle and ragworms. Despite active research in many other fields—most notably his work with the influential Trawling Commission and important studies of the early life histories of fishes—it was the compilation of his monumental compendium of marine annelid worms that occupied McIntosh for most of his long and active life.

In the preface to the first volume of the monograph, McIntosh explained that marine annelids represented "a department of native Zoology which more than any other required investigation, and whose neglected condition formed the author's chief inducement to attempt something for its improvement." And in the preface to the final volume, published fifty years later, he modestly acknowledged that for these worms, "there are many gaps to fill in literature, anatomy, physiology and development, but he hopes that they are left in a better state than he found them" and that

1. This magnificent king ragworm (*Alitta virens*) was found by McIntosh on a beach near St. Andrews and was taken home live and given to his beloved sister, Roberta, who prepared this stunning image.

"at least the treatise will give a foundation to the work of those who may follow, and conserve the time and labour so readily absorbed in the study of the group."

McIntosh's monograph was a major scholarly achievement and it was his hope that the beauty of the accompanying illustrations, many of which were drawn from life, would assist in rescuing his worms from "the comparative obscurity in which they have hitherto been involved," thereby bringing them to the attention of a wider public. In their ornamentations and colors, he felt that these worms were among the most splendid of all invertebrate types and that they easily vied with the "gaudy tints of butterflies and birds or the burnished splendour of beetles."

It had been his sister, Roberta, who had been his companion and workmate on many of the early collecting trips along the shores of St. Andrews, and whose exquisite drawings had encouraged him to take on the work of describing the rich variety of forms and functions of the marine worms they found. After his sister's death, McIntosh was fortunate to find another artist, Miss A. H. Walker, whom he felt most ably took over the task of completing and skillfully adding to the beautiful drawings that had been begun by his sister. But it was to his beloved sister, whom he described as his fellow worker and artist, that he dedicated these magnificent volumes.

The segmented worms of the Phylum Annelida include such familiar animals as earthworms and leeches, but the majority of annelid worms are mostly marine, and known as polychaetes. Common names for this remarkably diverse group of animals are: bristleworms, lugworms, featherduster worms, and sea mice. Polychaete means "many bristles," and most of these worms bear numerous bristles (or *setae*) on series of segmentally arranged side flaps called parapods. The variety in shapes, sizes, colors, and lifestyles among the more than nine thousand species of polychaetes currently known is extraordinary, and they are found throughout the world, living in the ocean depths, free floating on the sea surface, gliding through rock pools, or burrowing in mud and sand at the seashore.

2. These colorful polychaete worms are members of the Family Cirratulidae. Most cirratulids live in burrows in mud and sediment, with just their heads and elongate feeding tentacles visible above the substrate.

3. Paddleworms (*Eulalia tripunctata* and *Eulalia viridis*) are active predatory polychaetes in the Family Phyllodocidae. Figured at top center is a cocoon full of fertilized eggs.

4. Also highly predatory are these colorful eunicids, or Bobbitt worms, some growing to nearly 3 meters (c. 10 feet). Unlike phyllodocids, eunicids are stealth hunters, spending time buried in wait of passing prey. Once located, the speed of their attack, combined with their formidable jaws, can slice their prey in two.

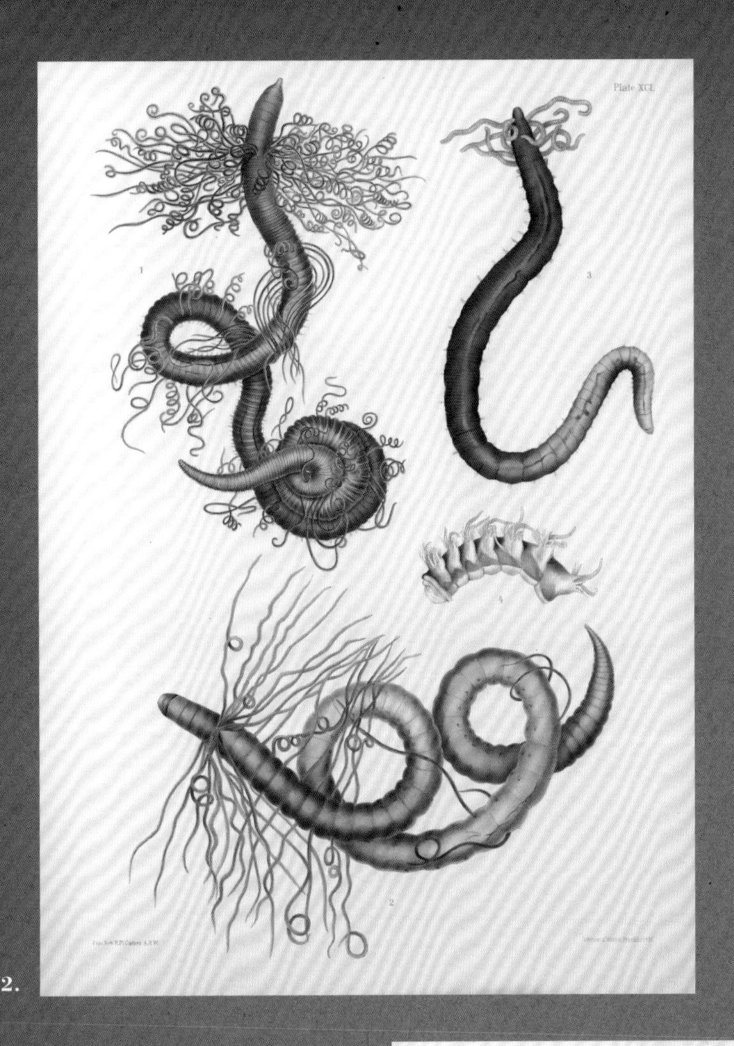

2.

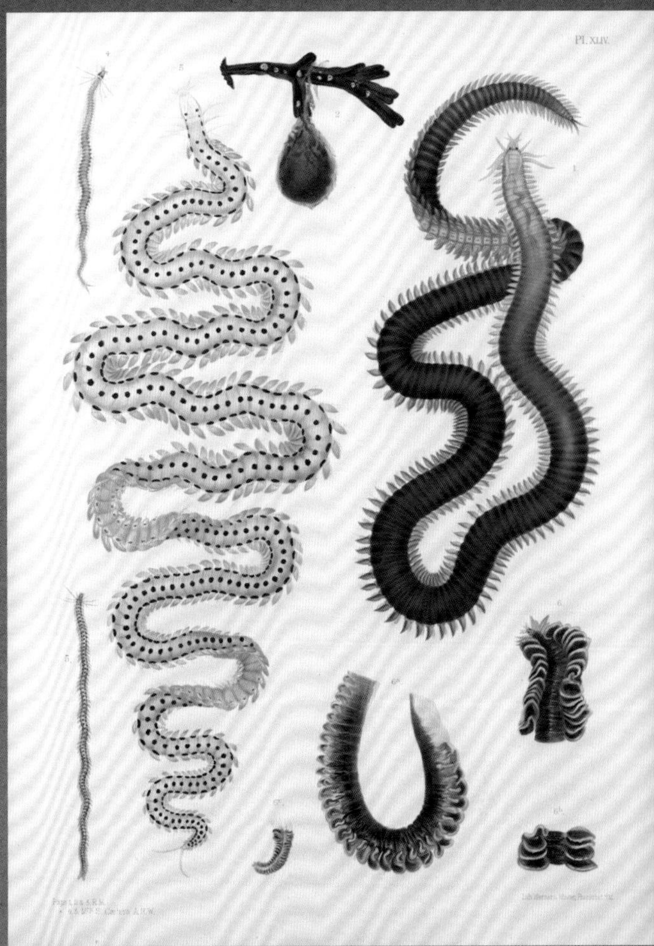

3.

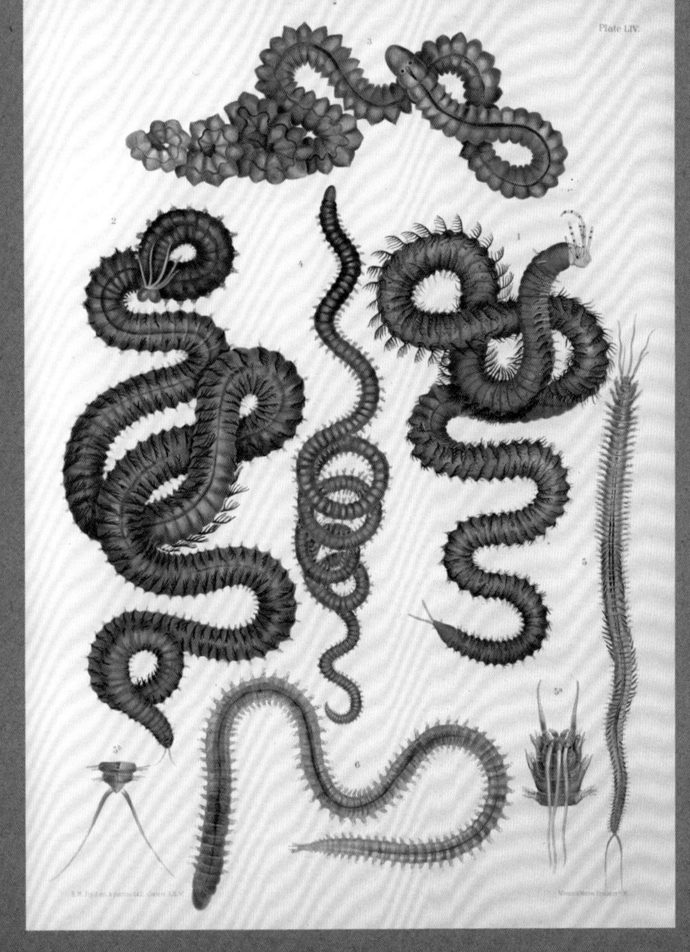

4.

The Hunter's Love for the Hunted

Author
Charles Melville Scammon (1825–1911)

Title
The marine mammals of the north-western coast of North America, described and illustrated: together with an account of the American whale-fishery

Imprint
San Francisco: J. H. Carmany; New York: G. P. Putnam's Sons, 1874

1. The Sulphurbottom whale, known today as the blue whale (*Balaenoptera musculus*), was too fast for the whalers of Scammon's time to capture with ease. The invention of the Foyn grenade harpoon initiated the wholesale slaughter of these magnificent animals in the late 1800s.

Charles Melville Scammon was born in Pittston, Maine, a small community on the shores of the Kennebec River about 50 kilometers (c. 30 miles) from the coast. His father was a town dignitary who served in various capacities, and later in the state legislature. The Scammon family was economically comfortable and a number of Charles's older siblings went on to achieve military and financial prominence in nineteenth century America.

Unlike his brothers, from a very early age Charles Scammon was entranced by the sea, and at the age of fifteen he petitioned his father to allow him to embark on a nautical career. At first his father would not agree to Charles leaving Pittston, but Scammon persisted and two years later he joined the crew of a Maine trading ship under the command of Captain Robert Murray. In 1842 he shipped out as Murray's apprentice and by 1848 he had assumed his first command. At the age of twenty-four, Scammon left the Maine coast as captain of a merchant barque and successfully completed the arduous voyage to San Francisco—a journey of over 168 days covering some 31,400 kilometers (c. 17,000 nautical miles).

Once in California, Scammon experienced firsthand the "gold fever" that had gripped the region, and soon realized that if he was going to be able to continue a life at sea, then he would have to "take command of a brig, bound on a sealing, sea-elephant, and whaling voyage." Thus it was that Scammon embarked on a career as a professional whaler, and he joined a fishery that was to see, within just a few decades, the decline to near extinction of many whale and seal populations.

By the mid-1850s, the American whaling fleet in the Pacific alone consisted of over 650 ships. This semi-industrialized fishery competed fiercely to provide the oil necessary for the lighting and heating of—and increasingly to lubricate the machinery for—a rapidly industrializing domestic market. While he was probably a whaler by necessity and not by choice, Scammon nonetheless proved to be a ruthlessly effective practitioner of the trade, and until his retirement from active hunting in 1863, he commanded numerous whaling and sealing vessels that traversed the Pacific hunting whales, elephant seals, and walruses.

He is perhaps best known for his fateful discovery of the gray whale calving lagoons of Baja California, where every year hundreds of pregnant gray whales would aggregate to give birth and nurse their calves, making them easy targets for his crews. Scammon's discoveries led to an eleven-winter intensive hunting effort from 1855 to 1865—known as the "bonanza period," when gray whaling along the California coast reached its peak. In the winter of 1859, Scammon sailed into San Ignacio Lagoon to the south of Laguna Ojo de Liebre (Scammon's Lagoon) and discovered the last remaining nursery grounds of the gray whales on the California coast. Within a couple of hunting seasons, that lagoon was almost completely devoid of whales.

Beyond his nautical skills and hunting prowess, Scammon is recognized today as a talented observer and a passionate—at times, even a jarringly compassionate—chronicler of the marine mammals he so ruthlessly hunted almost to extinction. This contradiction doesn't seem to have troubled him and he excitedly described scenes of the brutal wholesale slaughter of whales and seals as being "exceedingly picturesque," while at the same time meticulously documenting the habits, migration routes, diet, and anatomy of his quarry, even detailing hauntingly poignant observations of their strong parental and filial behaviors.

In culmination of many years of observation and detailed note-taking, in 1874 Scammon published his now famous treatise *The marine mammals of the north-western coast of North America, described and illustrated, together with an account of the American whale-fishery*. In the preface to this compendium, he noted the lack of scientific knowledge of even the most basic aspects of the natural history of these magnificent marine mammals, and observed that: "among the great number of intelligent men in command of whaling-ships, there was no one who had contributed anything of importance to the natural history of Cetaceans; while it was obvious that the opportunities offered for the study of their habits, to those practically engaged in the business of whaling, were greater than could possibly be enjoyed by persons not thus employed."

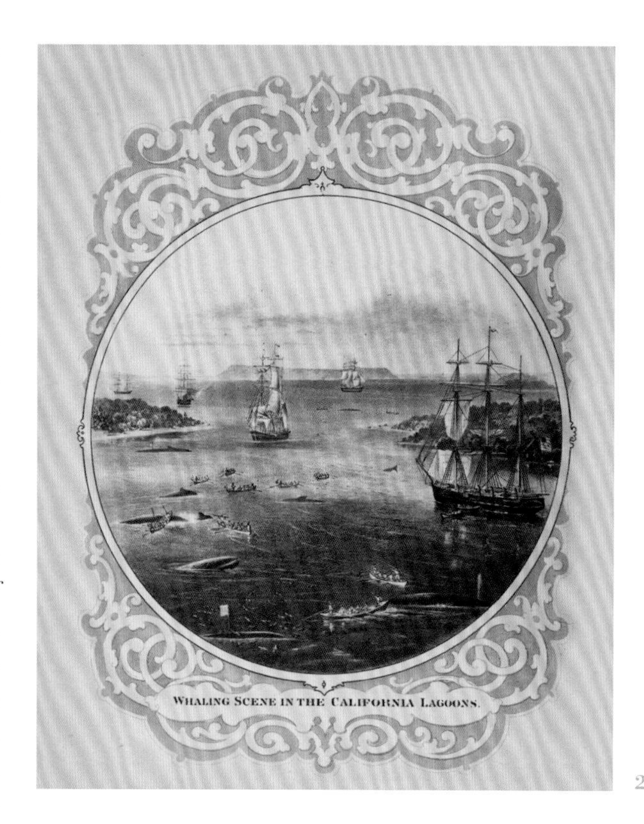

2.

2. Scammon's frontispiece depicts a scene of carnage of gray whales that had journeyed to the California lagoons to birth and suckle their calves. Commonly, whalers first killed the calf, knowing the mother would not leave it and could thus be easily harpooned.

To rectify this, Scammon set about the task of providing "correct figures of the different species of marine mammals found on the Pacific coast of North America" with "as full an account of the habits of these animals as practicable, together with such facts in reference to their geographical distribution as have come to my knowledge." He found "little difficulty in making satisfactory drawings of such smaller species of marine mammals as can be taken upon the deck of a vessel, but it is extremely difficult to delineate accurately the forms of the larger Cetaceans." It was not only that decomposition after capture was a problem, but also that the living animals "change their appearance in the most remarkable manner with every change of position, so that it is only from repeated measurements and sketches, and as the result of many comparisons, that I have been able to produce satisfactory illustrations of these monsters of the deep."

Scammon's book, which is written in "plain and simple language" and accompanied by many excellent illustrations, remains an invaluable reference, not just as a history of North American whaling, but also for the wealth of natural history data and observations he provided on the marine mammals of the Pacific Ocean. Most of the illustrations of whales and seals that accompany the volume were produced by the artist and lithographer Jacques Joseph Rey (1820–1892) of the famous Californian lithography firm of Britton and Rey and were based on Scammon's field drawings and measurements. For the depictions of sea and landscape backgrounds, Scammon acknowledged the skillful renditions made by Henry Steinegger (1831–1893), a junior partner in the lithography firm.

3. Pinnipeds, like cetaceans, are encased in thick layers of blubber. As whale numbers declined, hunters turned increasingly to large pinnipeds, such as the northern elephant seal (*Macrorhinus angustirostris*) and the stellar sea-lion (*Eumetopia jubatus*).

4. Steinegger's landscape features a group of gray whales (*Eschrichtius robustus*), a species that Scammon hunted to commercial extinction in California, in their feeding grounds among ice flows in the cold waters of the northeastern Pacific.

Although Scammon's work included accounts of seals, walruses, and sea otters, his central focus was on the fully aquatic cetaceans. The Cetacea is one of the most highly specialized of all mammalian orders, with about eighty living species divided into two main groups: the Mysticeti (baleen whales) and the Odontoceti (toothed whales, dolphins, and porpoises). Most cetaceans belong to the second group and have teeth on their jaws that are used for catching prey, although once caught, prey is not chewed but swallowed whole.

Some of the larger odontocetes, such as the Orca (*Orcinus orca*), will use their teeth to bite off chunks of large prey like seals or sea lions. The largest odontocete species is the sperm whale (*Physeter macrocephalus*), the males of which can grow to lengths of nearly 21 meters (c. 70 feet), although most males and all females are smaller. The sperm whale is a deep-diving species that feeds primarily on giant and colossal squid (see page 136).

The Mysticeti, on the other hand, lack teeth as adults, and instead have massive keratinous plates (baleens), which hang from their upper jaws and are used to filter food, usually krill and small fish, from the water column. Unusual among baleen whales, the gray whale (*Eschrichtius robustus*), with which Scammon's name is so closely associated, is a benthic feeder specialized to scoop up crustaceans from the seafloor.

The term "great whales," often used in older literature, usually refers to the thirteen largest cetacean species, ranging in size from the massive blue whale (*Balaenoptera musculus*), the females of which can reach lengths of almost 30 meters (c. 100 feet), to female common minke whales, which average around 10 meters (c. 32 feet) in length. It is interesting to note that Scammon was unable to provide much information on the blue whale, a species he, as had Herman Melville (1819–1891) in his classic *Moby-Dick*, termed the "sulphur-bottom whale," as this elusive animal was too fast for ready capture by the whaling boats of Scammon's time.

3.

4.

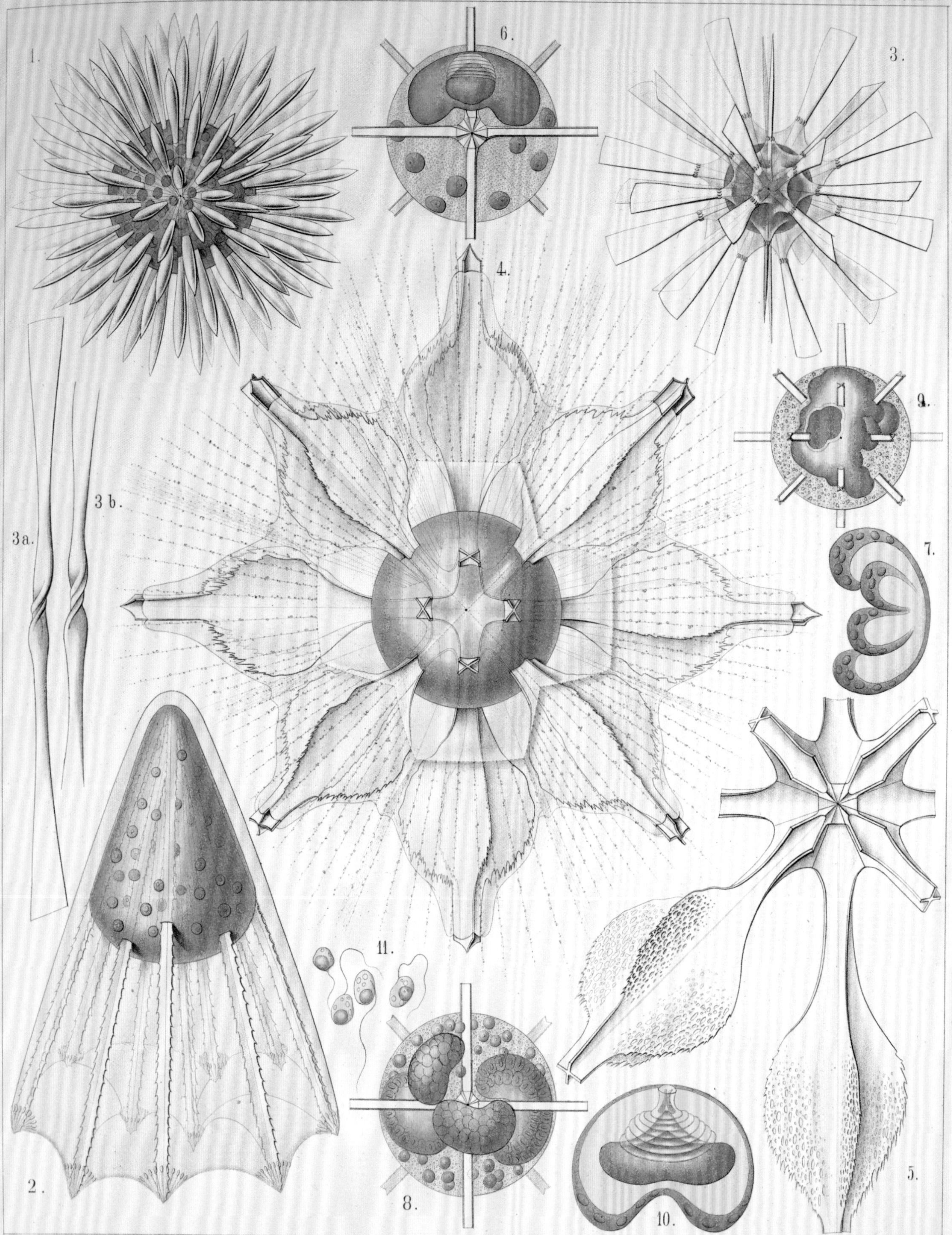

E.Haeckel and A.Giltsch,Del.

E.Giltsch Jena, Lithogr.

1. ACTINELIUS, 2. LITHOLOPHUS, 3. CHIASTOLUS,
4-11. ACANTHONIA.

Haeckel: Artist, Zoologist, and Evolutionist

Author
Ernst Haeckel (1834–1919)

Title
Report on the Radiolaria collected by H.M.S. Challenger during the years 1873–1876. (Report on the scientific results of the voyage of H.M.S. Challenger during the years 1873–1876. Zoology, v. 18.)

Imprint
London: Her Majesty's Stationery Office, 1887

1. The beautifully intricate and seeming endless variety of form that Haeckel observed in these microscopic radiolarians provided an outlet for both his scientific and artistic prowess.

The German zoologist Ernst Heinrich Phillipp August Haeckel is one of the great names of nineteenth-century biology. A prolific writer and talented artist, he became one of the most forceful and outspoken advocates for evolutionary ideas in continental Europe. He was also a great popularizer of science, and Haeckel's lectures and his books—which at the time sold many more copies in many more languages than did Darwin's own—played a critical role in promoting evolutionary thought to a wide European audience.

Haeckel was born to comfortable circumstances in Prussian Potsdam and, like so many privileged young men of his time, he studied medicine. After attending classes in Berlin and Würzburg, he obtained a license to practice in 1857 but within a year had decided that this was no career for him. He left for the Mediterranean and spent a year traveling, painting, and collecting marine life.

While at Messina in Sicily, he first encountered huge aggregations of the marine microorganisms known as radiolarians. Viewed under a microscope, their glassy (silica) skeletons revealed unimaginably intricate, often perfectly symmetrical geometric forms. For Haeckel, who was struggling to reconcile his idealist romanticist beliefs with the seeming reductionism of modern biology, these extraordinarily beautiful creatures provided a way to resolve the conflict. Study of their endless forms provided an outlet for his artistic talents while revealing to him nature's inner mysteries. Entranced, he returned to Germany and began doctorate studies at the University of Jena under the famed anatomist Karl Gegenbauer (1826–1903). In 1862 the first volume of Haeckel's *Die Radiolarien (Rhizopoda radiaria): eine Monographie* appeared in print and, that same year, he was appointed professor of comparative anatomy at the university, a position he was to hold for the next forty-seven years.

Haeckel was an accomplished anatomist and he worked on and beautifully illustrated a wide variety of marine invertebrates, ranging from microscopic monerans to large colonial siphonophores (see page 44). But it is with the Radiolaria that he continued to make his most scholarly contributions. While busy with his radiolarian monographs, Haeckel was contacted by Sir Charles Wyville Thomson (1830–1882) with an invitation to contribute a report on the radiolarians collected during the HMS *Challenger* expedition of 1872–1876. Wyville Thomson, the initiator and chief scientist on the *Challenger*, was engaged with the gargantuan task of finding experts to work up the huge collections of organisms that had been gathered during that expedition. Haeckel graciously accepted, and for the next ten years worked assiduously on the *Challenger*'s radiolarians. A lesser mind might have been overwhelmed by the enormity of the task—to describe and delineate radiolarian forms as innumerable as "the boundless firmament of stars." But Haeckel persisted and finally in 1887 published his monumental *Report on the Radiolaria collected by H.M.S. Challenger* as the eighteenth volume of the zoological reports of the expedition.

His monograph appeared in three parts, the third being an atlas of some sixteen hundred exquisitely rendered lithographs. In many ways, it was Haeckel's stunning drawings that served to stimulate widespread appreciation for "this inexhaustible kingdom of microscopic life, whose endless variety of wonderful forms justifies the saying—Natura in minimis maxima." Their diversity had captivated and inspired Haeckel, and in turn his artistic talents brought a whole universe of microscopic life to the world's attention. In this extraordinary work, Haeckel described 739 genera and 4,318 radiolarian species, of which 3,508 were new to science. Even so, he was acutely aware that he had far from exhausted the diversity at hand, acknowledging that "a careful and patient worker who would devote a second decade to the work, would probably increase the number of new forms (especially of the smaller ones) by more than a thousand; but for a really complete examination, the lifetime of one man would not suffice."

After Darwin, Haeckel was probably the most influential evolutionist of the nineteenth century. He wrote prolifically and, in addition to his scientific works, he published many widely read books championing evolutionary ideas to a lay audience. Although a vociferous supporter of Darwin, Haeckel never believed natural selection to be the main driver of evolution, and instead developed alternative ideas. He believed that evolution progressed in response to environmental influence directly through changes in embryonic development, and that "ontogeny recapitulates phylogeny," ideas that are today considered somewhat flawed. Nonetheless, despite the controversial nature of some of his theories, Haeckel was a profound thinker, and his works have bequeathed to the field of biology some of its fundamental modern concepts, coining such enduring terms as ecology, phylogeny, and phylum, among many others.

Radiolarians are planktonic, unicellular marine eukaryotes (organisms with a membrane-bound nucleus) found in all of the world's oceans, where they have been an important component of marine ecosystems for the past 500 million years. Despite being single celled, radiolarians are anatomically complex organisms, with multiple body compartments supported and surrounded by an elaborately constructed skeleton known as a test. The most striking features of these organisms are their often highly elaborate siliceous tests, which may be formed of complex lattice of plates or large numbers of sharp spicules through which threadlike extensions (rhizopodia) of the inner cell's cytoplasm extend. Because silica is rather resistant to dissolution in seawater, their skeletal remains sink and accumulate, forming much of the siliceous ooze that covers large areas of the seafloor.

Many of the radiolarians described by Haeckel had been dredged up in the deep-sea oozes collected by the *Challenger* expedition. Radiolarians feed on other plankton such as diatoms, algae, copepods (see page 128), and bacteria, which they entrap using their numerous sticky rhizopodia. Many species also house symbiotic photosynthetic algae (mainly dinoflagellates) in their outer body compartments; these produce nutrients for them and in turn the algae are provided protection and supplied with nutrients in the form of their hosts' nitrogenous and carbon dioxide wastes.

2. Here, Haeckel illustrates eleven species of the radiolarian genus *Hexastylis*. The animal in the center of the plate is the beautiful radiolarian *Hexastylis cochleatus*, named in reference to its striking red coloration.

3. In 1879 John Murray erected the genus *Haeckeliana* in Haeckel's honor, and the radiolarian in the center of this plate is a species named *Haeckeliana darwini* by Haeckel in tribute to his intellectual mentor.

4. Haeckel considered *Gorgonetta*, a genus he described in the *Challenger* report, among the most remarkable of all radiolarian forms. *Gorgonetta mirabilis*, meaning amazing or wondrous, was the name he chose for this species.

5. Many radiolarians, particularly those found in sunlit layers of the oceans, contain symbiotic (green) algae in their tissues. These algae provide nourishment for their hosts while receiving protection from predation.

The Voyage of H.M.S. "Challenger."
Radiolaria Pl. 21
HEXASTYLUS.

2.

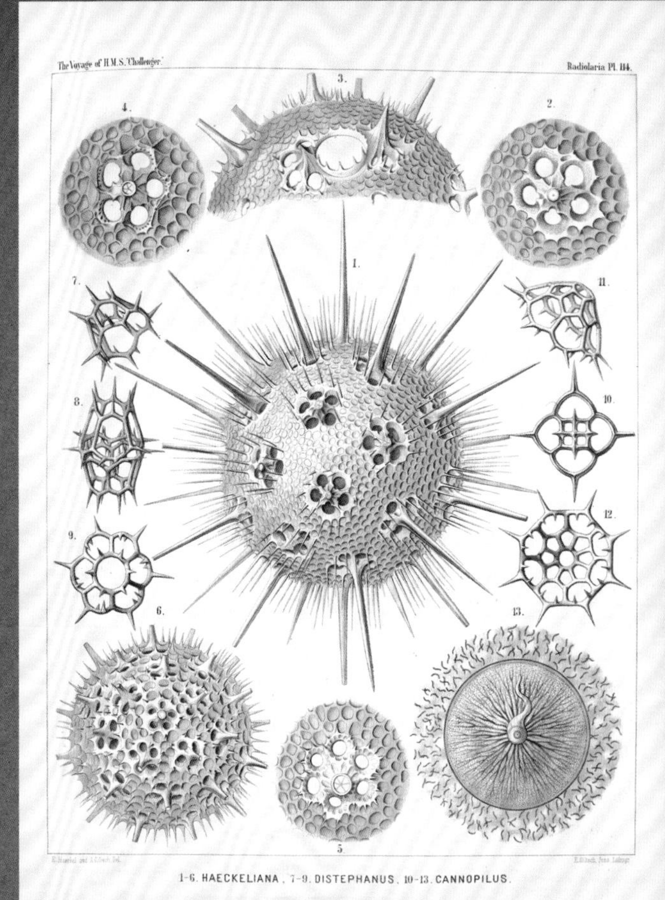
The Voyage of H.M.S. "Challenger."
Radiolaria Pl. 114.
1-6. HAECKELIANA, 7-9. DISTEPHANUS, 10-13. CANNOPILUS.

3.

The Voyage of H.M.S. "Challenger."
Radiolaria Pl. 119.
GORGONETTA.

4.

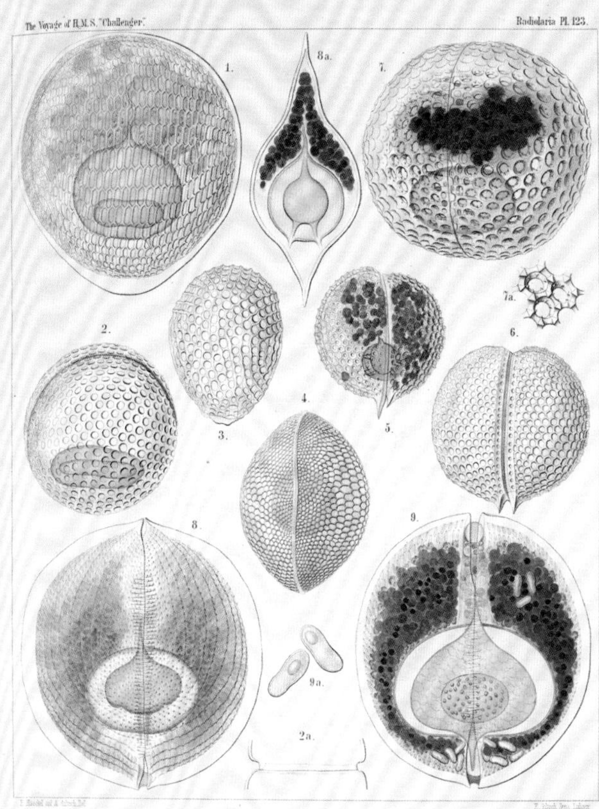
The Voyage of H.M.S. "Challenger."
Radiolaria Pl. 123.
1-4. CONCHARIUM, 5,6. CONCHASMA, 7. CONCHELLIUM,
8,9. CONCHOPSIS.

5.

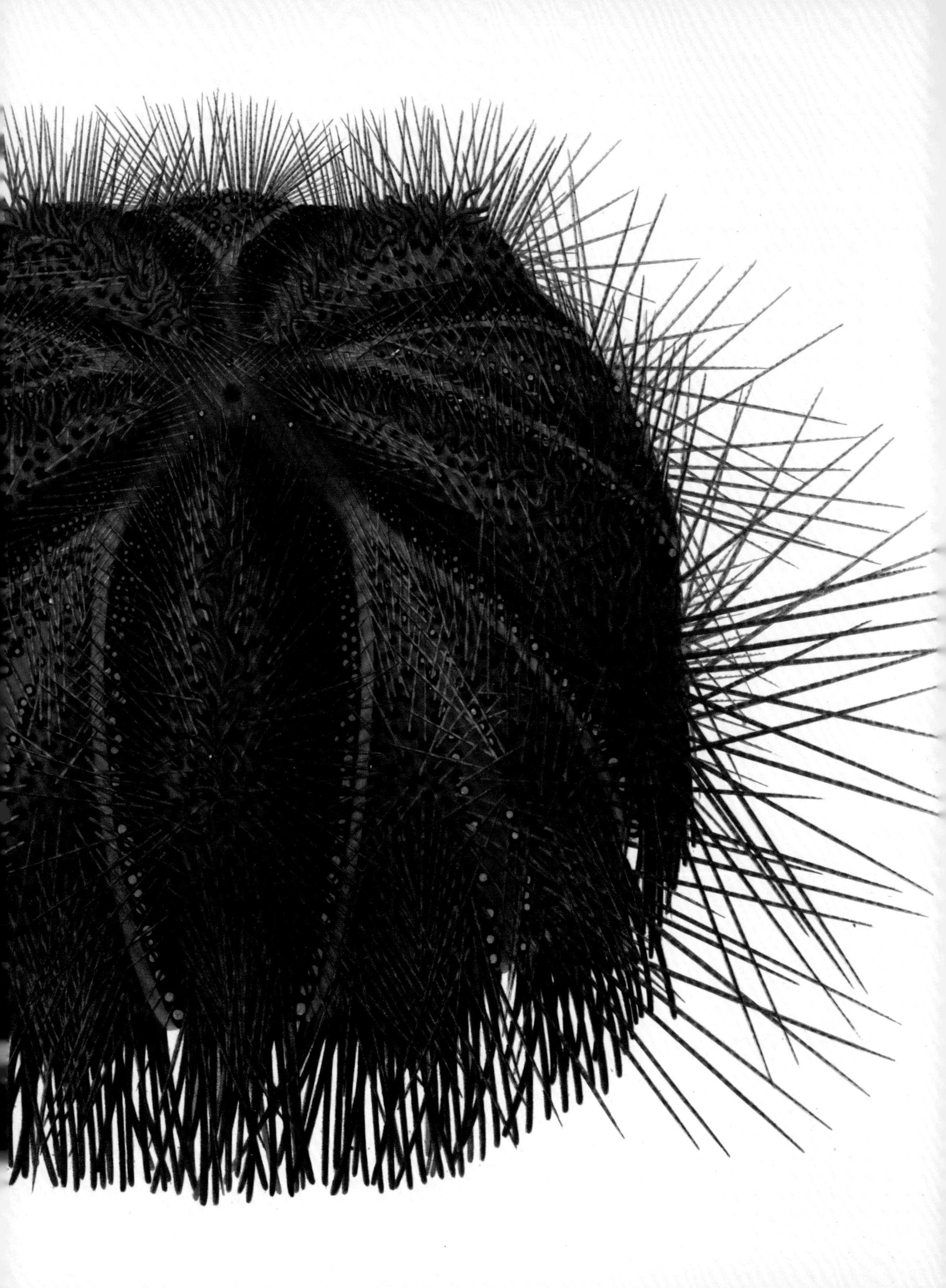

A Pioneer in Conservation

Authors
Paul Sarasin (1856–1929)
Fritz Sarasin (1859–1942)

Title
Ergebnisse naturwissenschaftlicher Forschungen auf Ceylon

(Results of scientific research in Ceylon)

Imprint
Wiesbaden: C. W. Kreidel, 1887–1893

1. The blue-spotted fire urchin (*Astropyga radiata*) is a common Indo-Pacific species. The talented Sinhalese artist William de Alwis drew this beautiful specimen for the Sarasins during their stay at Peradeniya.

The biologist Paul Benedikt Sarasin was the founder of the Swiss nature conservation movement and a forceful advocate for international conservation at the turn of twentieth-century Europe. Born and raised in Basel, he began medical studies there, but after visiting the Zoological Institute in Würzberg, he fell in the thrall of the charismatic German explorer and naturalist Karl Gottfried Semper (1832–1893) and soon abandoned medicine, turning instead to zoology. Under Semper's mentorship, Sarasin completed a thesis on the development of the European faucet snail (*Bithynia tentaculata*), for which he received his doctorate from the University of Würzburg in 1882.

In the fall of the following year, Sarasin left Europe, accompanied by his second cousin, Fritz (Karl Friedrich Sarasin, 1859–1942), on a two-and-a-half-year expedition to the then British-held Indian Ocean island of Ceylon (Sri Lanka). The cousins embarked on an ambitious undertaking traversing the length and breadth of the island, making collections and observations of plant and animal life, detailed geographic observations, and extensive ethnological studies of the island's indigenous peoples.

Returning to Basel in the spring of 1886, they began work assembling their findings into the impressive three-volume *Ergebnisse naturwissenschaftlicher Forschungen auf Ceylon*, which was published between 1887 and 1893.

The work is beautifully illustrated and, while Sarasin makes no reference to the artists who produced the plates, a number of the illustrations, such as those accompanying this essay, bear the name "de Alwis" in the lower left-hand corner. It is known that the de Alwises were a family of Sinhalese artists based at the botanical gardens at Peradeniya from the early 1820s. In the introduction to the *Ergebnisse*, Sarasin mentions that he and his cousin made an extended stay at Peradeniya during their second year on the island.

Harmanis de Alwis (1792–1894) was a renowned botanical illustrator and his two sons were also accomplished artists. The elder of the sons, William de Alwis (1842–1916), was also a highly regarded botanical illustrator, but is known to have made zoological drawings for naturalists who visited the gardens, and there is little doubt that Sarasin's beautifully rendered sea urchins are the work of this talented Sinhalese artist.

Following the publication of *Ergebnisse*, the Sarasin cousins traveled to the Far East and spent almost three years exploring the Celebes (Sulawesi). Returning in 1896, they published a report on their findings there.

In the following years, Paul Benedikt Sarasin expanded his interests and wrote extensively on subjects ranging from zoology and ethnology to astronomy, art history, and theology. By the turn of the century, Sarasin had become committed to the cause of nature conservation, and at the 1906 annual meeting of the Swiss Society for Nature Research, he successfully lobbied for the creation of a conservation commission. In 1909 he founded the Swiss League for the Protection of Nature. Due in no small part to the efforts of Sarasin and the league, in 1910 the foundation stone was laid in Switzerland's beautiful Val Cluozza Valley in what was to become the first national park in central Europe.

Although successful in his native Switzerland, Sarasin realized that an international organization was needed to coordinate conservation efforts worldwide. Working tirelessly on the international stage, Sarasin succeeded in 1913 in bringing together representatives of sixteen European nations and the United States to establish a Consultative Commission for the International Protection of Nature. Sadly, conflict and inaction soon led to the dissolution of the commission, and Sarasin's declining health forced him to retreat from public life.

In the spring of 1929, he succumbed to pneumonia and died in his hometown of Basel at the age of seventy-three. While Sarasin did not live to see his dream of an international conservation network realized in his lifetime, his vision is credited with the formation in 1948 of the International Union for Conservation of Nature (IUCN), and he is today honored as a pioneering statesman of the international conservation movement.

Sea urchins are members of the large Phylum Echinodermata and together with sand dollars are grouped in the Class Echinoidea. They are highly successful marine invertebrates, with about seven hundred species known throughout the world's oceans, where they live in habitats from the intertidal zone to the deep ocean. Some species aggregate in groups of hundreds or even thousands of individuals, while others are solitary. They vary widely in body shape, ornamentation, and coloring, yet most are immediately recognizable by their rounded symmetrical outer test, variously domed dorsally and flattened on the ventral surface, and usually bearing protective spines of some sort. Close examination reveals the fivefold symmetry characteristic of most echinoderms, and even if not obvious externally when dissected, as in the accompanying figure, all internal organs adhere to this fivefold pattern.

Sea urchins are able to move, albeit far more slowly than the related brittle stars (see page 26), by using five paired rows of so-called tube feet, which penetrate the ventral surface of the test and are internally connected to the peculiar water vascular system unique to echinoderms, which serves in locomotion as well as respiration. Valued in many parts of the world for their gonads, both male and female gonads are considered a delicacy and, as a result, they are heavily overfished in some regions.

Loss of sea urchins, either from overfishing or from increasingly frequent disease outbreaks, is devastating to reef communities because in the absence of these grazing herbivores, reefs can become chocked with algae and die.

2. & 3. When dissected open, the five-fold symmetry, so characteristic of most echinoderm species, is clearly evident in all of the internal organs of this urchin.

4. The short spines of this beautiful Indo-Pacific fire urchin (*Asthenosoma varium*) are venom-tipped and can inflict a very painful sting that serves as an effective antipredation device.

Tafel XIV.

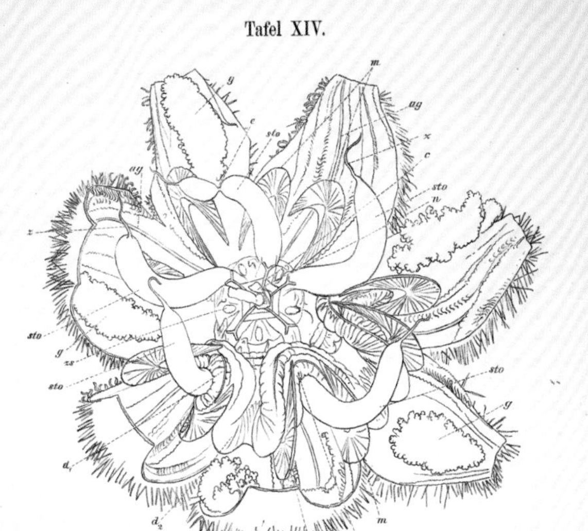

Fig. 27. Die Stewart'schen Organe von Asthenosoma (Seite 100 ff.)

ag Ambulacralgefäss, *c* Compassstücke der Laterne, d_1 erste untere, d_2 zweite obere Darmwindung, *e* Einschnürung einer Stewart'schen Blase, *g* Geschlechtsorgane, *m* Längsmuskeln, *n* Niere, *sto* Stewart'sche Organe, *z* zipfelförmiger Anhang derselben, *zs* Zahnsäcke der Laterne.

2.

27.

Mützel del.

Lith. Anst. v. Werner & Winter, Frankfurt a/M.

C. W. Kreidel's Verlag, Wiesbaden.

3.

SARASIN, Ceylon Bd. I.

Taf. X.

4.

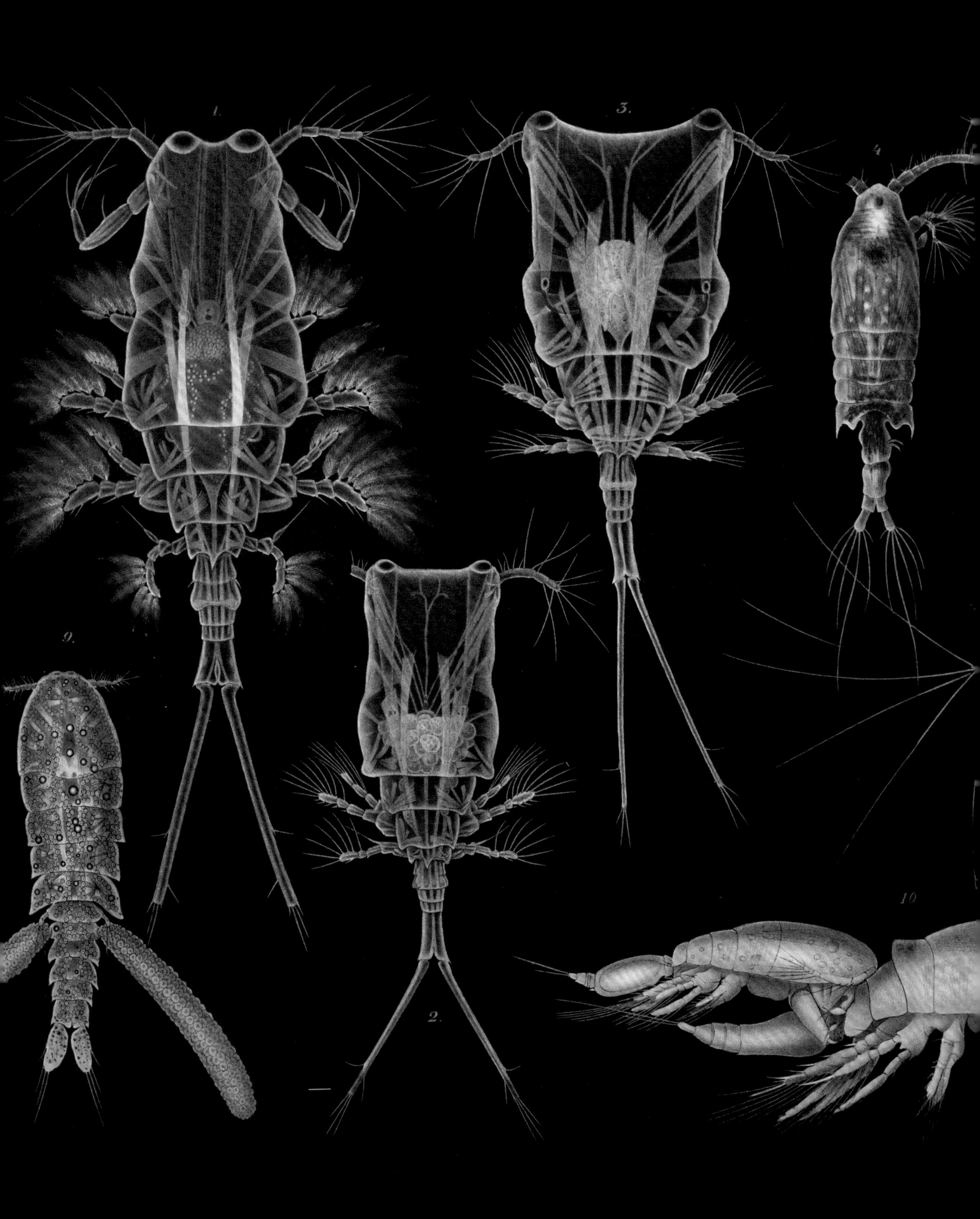
1.
3.
4
9.
2.
10

The Golden Age of Copepodology

Author
Wilhelm Giesbrecht
(1854–1913)

Title
Systematik und Faunistik der pelagischen Copepoden des Golfes von Neapel und der angrenzenden Meeres-Abschnitte. (Fauna und Flora des Golfes von Neape und der angrenzenden Meeres-Abschnitte, 19. Monographie.)

(Systematics and faunistics of pelagic copepods of the Gulf of Naples and its adjacent marine regions. [Fauna and flora of the Gulf of Naples and its adjacent marine regions, 19th monograph.])

Imprint
Berlin: R. Friedländer & Sohn, 1892

1. Giesbrecht was among the first to document bioluminescence in copepods, and many of his marvelously detailed plates are set against a black background to illustrate their glowing colors.

The period around the turn of the nineteenth century known as the Golden Age of Copepodology was a time of rapid expansion in knowledge of these tiny marine crustaceans. It was during this time that marine biologists began to recognize the central role that these abundant organisms play in the ocean ecosystem. One of the great copepod specialists active during the golden age was the German zoologist Wilhelm Giesbrecht, and his stunningly illustrated monograph *Pelagische Copepoden* is today considered one of the truly seminal works of the period.

Giesbrecht was born in Danzig, then part of the Prussian Empire, and began his studies in 1878 at the Zoological Institute of the University of Kiel. The Zoological Institute had been founded ten years earlier with the famous German marine ecologist Karl August Möbius (1825–1908) as its influential first director. Under Möbius's leadership, the Zoological Institute had rapidly grown into a center of academic excellence, attracting students and faculty from throughout the Prussian Empire. The institute was to become the birthplace of ecological science in northern Europe, and Möbius's works on the fauna and ecology of the Bay of Kiel laid a foundation for the science of marine ecology. His influence on the many students who passed through Keil was profound.

It was under Möbius's direct supervision in 1881 that Giesbrecht completed his thesis on the copepods of the Baltic region. In the following years, Möbius began a series of influential studies of the role of copepods in the oceans, but by that time, Giesbrecht had left for a much-coveted position as a staff zoologist at the world-famous Zoological Station in Naples. The station's founder, Felix Anton Dorhn (1840–1909), held Giesbrecht in high regard and appointed him one of only seven on the scientific staff at the station.

In Naples, Giesbrecht had access to the world-class facilities and intellectual capital of that international scientific hub, and was able to assemble huge collections of copepod specimens from the bay and surrounding regions. His monographic studies benefited greatly from the addition of the copepod collections of Lieutenant Gaetano Chierchia (1850–1922) made during the 1882–1885 circumglobal expedition of the Italian corvette *Vettor Pisani*. Dorhn had trained Chierchia a few years earlier and, although the main scientific goals were surveying, deep-sea sounding, and thermometric measurement, Chierchia and other officers had made collections of marine organisms from around the globe.

After more than ten years of study, Giesbrecht published his monumental *Pelagische Copepoden (Systematik und Faunistic)* in 1892. Among the fifty-four lithographic plates that accompany the volume are some extraordinarily beautiful colorized depictions of copepods set against a black background—invoking their dark, deep-water habitats. One such copepod species illustrated is *Oncaea conifer*, a species discovered and described by Giesbrecht. He beautifully illustrated a small male copulating with a large female, and just a few years later, published the first observations of bioluminescence (see page 140) in the species. Twelve years after the publication of his monograph, Giesbrecht was appointed honorary professor at the Zoological Station and spent the rest of his career working there. He died in Naples in February 1913 at the age of fifty-nine.

Copepods are an important group of aquatic crustaceans. Over thirteen thousand species have been described and at least ten thousand of these live in the oceans, where they are found in every conceivable habitat, from the surface waters to the deepest abyss. Although most are small, typically only 0.2 to 2 millimeters in length, they are among the most abundant of all multicellular animals on the planet. About half of all copepod species are parasites living either inside or on fish and many invertebrates, but the rest are free living and it is these pelagic forms that make up the great bulk of the ocean's zooplankton.

Copepods form an important link in the ocean food web because they are the main predators of the unicellular bacteria, protists, and marine algae that convert the sun's energy into edible biomass. In turn, copepods are eaten by innumerable other invertebrates, fish, and even by whales and seabirds. Copepods also perform another very important role in the oceans, and that is in the cycling of carbon. Many of them feed in the surface waters of the ocean at night, and during the day they migrate down into deeper waters to avoid predation. In this way, they carry carbon produced at the sunlit surface into the deeper sunless zones.

Their feces, molted outer cuticles, and dead bodies also rain down from the surface waters, forming a large component of what has been termed "marine snow." The continuous shower of that organic detritus carries carbon down to the deep ocean floor where it can remain "locked up" and out of contact with the atmosphere for hundreds of years.

2. In this remarkable plate, Giesbrecht depicts species against dark and light backgrounds. One, *Sapphirina ovatolanceolata* (bottom center) is drawn twice, once on a black background to highlight its beautiful bioluminescent colors, and again on a light background to feature details of its anatomy.

3. Giesbrecht's illustrations provide not only remarkably insight into copepod diversity but also details of reproductive biology. In this plate he illustrates some of the many ways females of different species carry fertilized eggs in extruded egg sacs, or attached to their bodies.

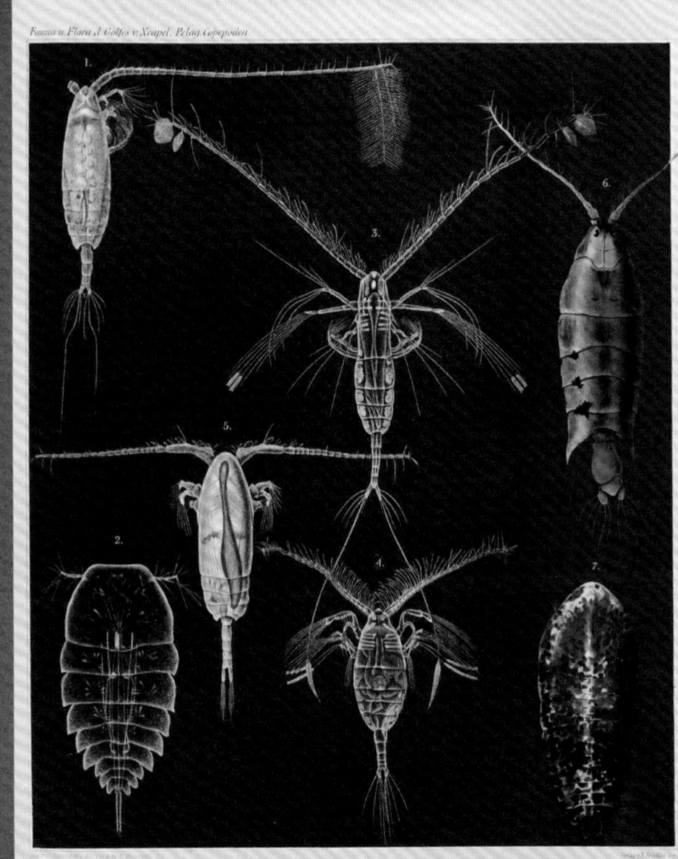
Fauna u. Flora d. Golfes v. Neapel. Pelag. Copepoden

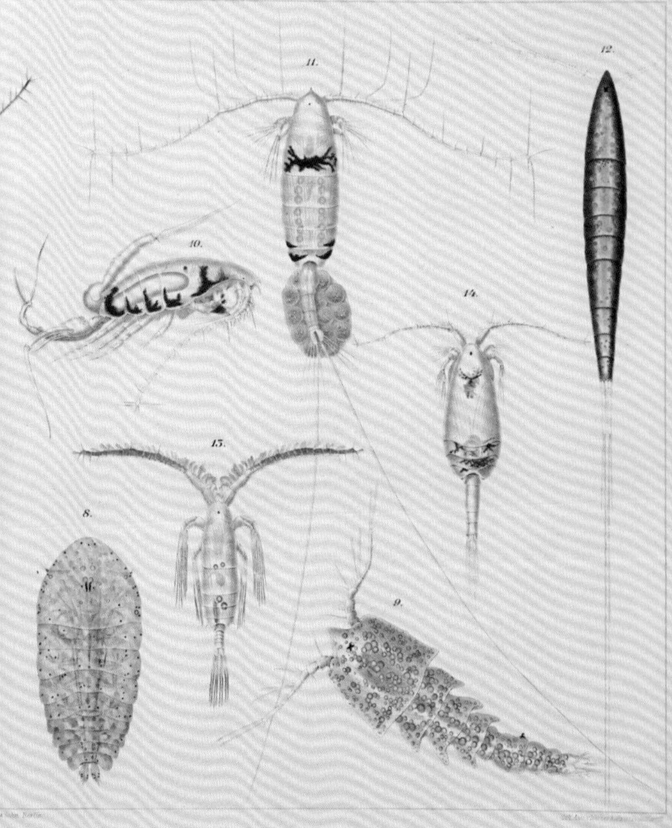
Taf. 1.

2.

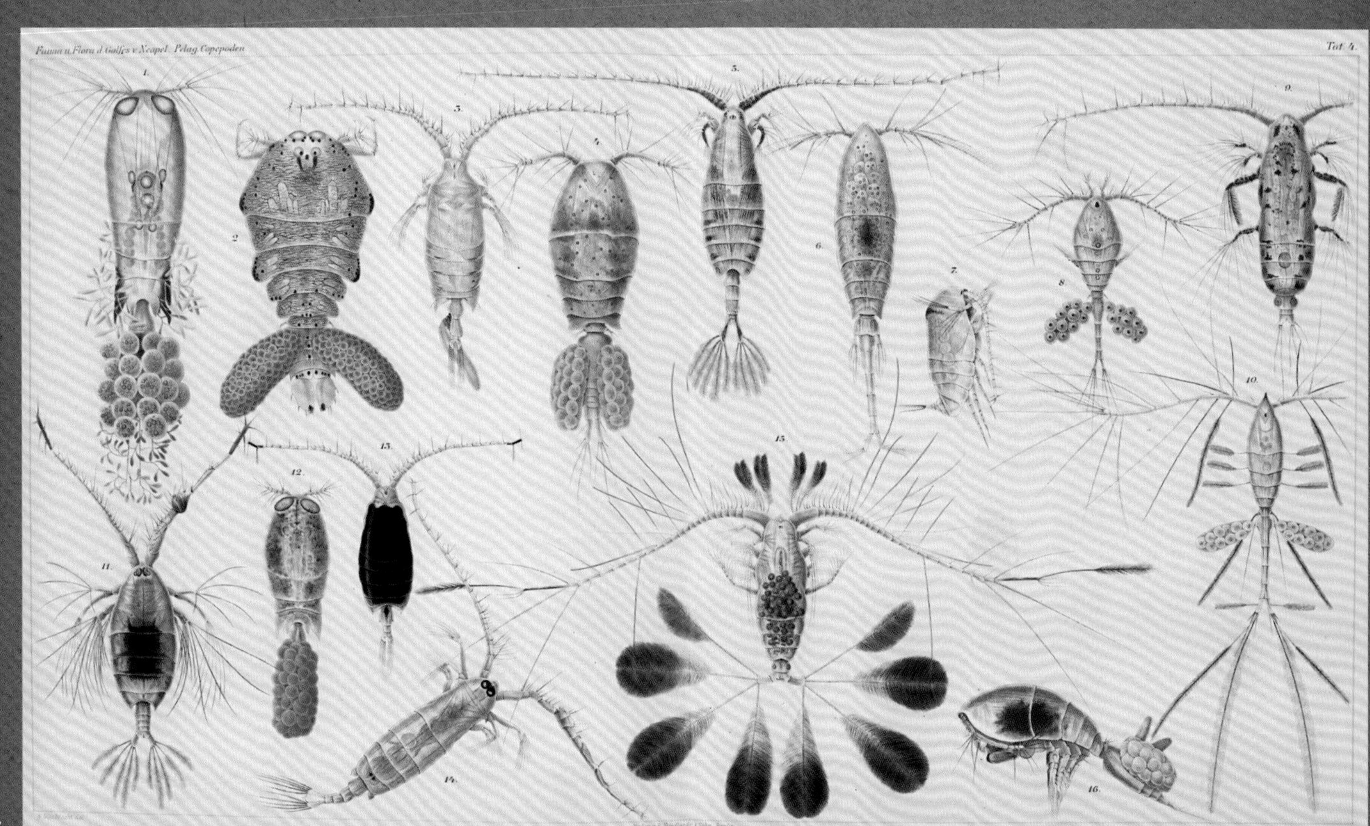
Fauna u. Flora d. Golfes v. Neapel. Pelag. Copepoden
Taf. 4.

3.

13.
14.
12.
11.
8.
m
15.
10.
9.
13.a

The Enigmatic Nemertean Worms

Author
Otto Bürger (1865–1945)

Title
Die Nemertinen des Golfes von Neapel und der Angrenzenden Meeres-Abschnitte. (Fauna und Flora des Golfes von Neape und der angrenzenden Meeres-Abschnitte, 22 Monographie.)

(The nemerteans of the Gulf of Naples and adjacent regions. [Fauna and flora of the Gulf of Naples and its adjacent marine regions, 22nd monograph.])

Imprint
Berlin: R. Friedländer & Sohn, 1895

Figures 3 & 4
Author: William Carmichael McIntosh (1838–1931)
Title: *A monograph of the British marine annelids*
Imprint: London: Ray society, 1873–1923

1. The bright colors and striking patterning of many nemertean worms is aposematic, warning predators of their unpalatability. The main predators of these nemerteans are other nemerteans, which appear tolerant of the toxins.

Willhelm Heinrich Otto Bürger was born and raised in Hanover in lower Saxony. Unfortunately, very little has been recorded about his family or his childhood, but it is known that by his midtwenties he was a lecturer in zoology and assistant at the Zoological Institute of the University of Göttingen. At Göttingen, Bürger established his reputation as a researcher on marine worms, particularly the strange ribbon, or proboscis, worms of the Phylum Nemertea. His mentor, and the world expert on nemerteans at the time, was the eminent Dutch embryologist Ambrosius Hubrecht (1853–1915). When Hubrecht invited Bürger to come to the world-famous Zoological Station in Naples to take over his nemertean work there, Bürger jumped at the opportunity.

Hubrecht had been working at the station in Naples for many years, but his interests were widening and an opportunity had arisen for him to travel to the Dutch East Indies, so he asked Bürger to complete a monographic study of the nemertean worms of the Naples region. With financial support from the Berlin Academy of Sciences, Bürger arrived in Naples in the winter of 1891 and immediately began work sorting through all of Hubrecht's materials and beginning his own studies. By the fall of 1893, Bürger had assembled all of the data necessary for the embryological and biological aspects of the study, but he had to return to his duties in Göttingen before he could complete the work.

In Göttingen, he continued working on the systematic sections of the monograph, and finally completed the entire manuscript, totaling over seven hundred pages, which was published in 1895 as *Die Nemertinen des Golfes von Neapel und der Angrenzenden Meeres-Abschnitte.* The work was accompanied by thirty-one lithographic plates, with many illustrations of nemertean worms rendered in the brilliant colors exhibited by these animals in life. Bürger very gratefully acknowledged the artist who prepared these stunning figures in the preface to the monograph. Sadly, it appears that the young man, named only as "Heinze," worked for a full six months carefully preparing the illustrations under Bürger's close supervision but died very soon after the drawings were completed, while Bürger returned to Germany.

Bürger's nemertean monograph was well received and considered one of the most important contributions on the enigmatic nemertean worms that had appeared up until that time. In 1900 Bürger left Göttingen for Chile, where he became director of the Chilean National Museum of Natural History in Santiago. In the previous year, the departments of botany, zoology, and mineralogy had been established at the museum, and Bürger was jointly appointed as a professor of zoology. He continued to work on nemertean worms, although administrative duties and travels around the country appear to have occupied most of his time. In 1908 he stepped down as director of the museum and returned to Germany. Although he continued to publish a few scientific papers, his main focus turned to economic geography and travel writings about his time in Chile. He died at the age of eighty in the Upper Bavarian town of Törwang.

At the time that Bürger was undertaking his nemertean studies, there had arisen quite a controversy regarding the relationships of these peculiar worms. Most zoologists thought them closely related to platyhelminth flatworms, but others such as William McIntosh (see page 111) considered them to be an offshoot of the polychaete annelid worms; the illustration of *Lineus longissimus*, a nemertean that is reported to reach a staggering 30.5 meters (c. 100 feet) in length, is taken from McIntosh's annelid monograph. Finally, and certainly most controversially, Bürger's mentor, Ambrosius Hubrecht, championed the idea that nemertean worms had given rise to the chordates (the precursors of all animals with backbones). Bürger weighed in on this controversy, and after a careful analysis of the pros and cons of each competing theory, came to the conclusion that nemerteans were probably more closely related to platyhelminth flatworms than to any other group. His position held sway for most of the following century.

Interestingly, today their relationships are again in doubt and, although no support has been found for a relationship with chordates, molecular studies suggest they belong to a large invertebrate group, the Lophotrochozoa, which includes annelids, platyhelminths, mollusks, brachiopods, and a range of other phyla—but no consensus has yet been reached as to where exactly the nemerteans fit.

Regardless of their relationships, approximately one thousand species of nemertean worms are known today, and the great majority of them are found in ocean habitats, where they burrow in seafloor sediments or in crevices between shells and rocks. While some can reach extraordinary lengths, often many scores of feet, most are less than 25 centimeters (c. 10 inches) in length, although all have highly extensible bodies capable of stretching many times longer than their resting lengths. Most move slowly over a slime produced by glands on the head, and use banks of external cilia to glide over the mucous trail. Some larger species are capable of movement propelled by waves of muscular contractions that allow them to crawl or swim through the water column. Most are active predators, feeding on a variety of prey such as annelid worms, mollusks and crustaceans, and even other nemerteans.

Prey capture is accomplished using a massively extensible proboscis, which is an enfolding of the body wall located above the digestive canal in a large muscle-lined cavity (the rhynchocoel). When body muscles compress the fluid in the rhynchocoel, the proboscis is rapidly everted and wraps around the prey, which is then immobilized by sticky toxic secretions or repeatedly stabbed with a sharp calcareous barb that injects toxins into the prey's body. Once the prey is immobilized, it is drawn into the nemertean's mouth as the proboscis is retracted.

2. Over one hundred species are currently included in the nemertean genus *Lineus*, and in this plate seventeen species of these colorful bootlace worms are artfully arrayed.

3. This attractive *Carinella annulata* is from the classic work of William McIntosh, who considered nemerteans to be an offshoot of polychaete annelids and, consequently, included many nemerteans in his monograph of British annelids.

4. Also from McIntosh, this arresting image is of the giant bootlace worm *Lineus marinus*, which is among the longest of all animals. Although nemerteans are notoriously hard to accurately measure, individuals of *Lineus marinus* of 54 meters (c. 177 feet) in length have been reported.

2.

3.

4.

The Success of the *Valdivia*

Author
Carl Chun (1852–1914)

Title
Aus den Tiefen des Weltmeeres: Schilderungen von der Deutschen Tiefsee-Expedition

(From the depths of the world's oceans: description of the German Deep-Sea Expedition)

Imprint
Jena: Gustav Fischer, 1900

Figures 1, 3, 4 & 5
Author: Carl Chun (1852–1914)
Title: *Die Cephalopoden (Wissenschaftliche Ergebnisse der Deutschen Tiefsee-Expedition, auf dem Dampfer "Valdivia," 1898–1899, Bd. 18)*
Imprint: Jena: G. Fischer, 1910–1915

1. The artist-engraver Fritz Winter prepared most plates for the twenty-four volume reports of the *Valdivia*. This magnificent specimen of the Antarctic octopus *Benthoctopus levis* beautifully illustrates his artistic and technical prowess.

The German zoologist Carl Friedrich Chun was a visionary of extraordinary talent and drive. He studied at the universities of Göttingen and Leipzig, and after receiving his doctorate in 1876, he moved to the Zoological Station in Naples, which at the time was considered the premier center for marine biological research. While in Naples, Chun published a monograph on ctenophores (see page 49), a work that brought him international acclaim and, in 1881, he was elected to the German Academy of Sciences Leopoldina.

Appointed as professor of zoology at the University of Breslau in 1891, Chun expanded his interests to include other planktonic (floating) invertebrates as well as cephalopod mollusks. A series of well-received publications established him as a recognized authority on marine biology.

In 1897 Chun delivered an impassioned address to the German Society of Naturalists and Physicians, outlining plans for an ambitious expedition to traverse the globe collecting invaluable data and specimens from the deep oceans. Prior to Chun's proposal, it had been primarily the British whose expeditionary acumen had dominated exploration in the deep seas, most notably with the epoch-making *Challenger* expedition of 1872–1876 (see page 119). Chun was convinced that much remained to be discovered in the ocean's depths, and that a full-equipped scientific expedition targeting areas neglected by the *Challenger* should be a national priority. His visionary proposal was well received and almost immediately gained the support of Emperor Wilhelm I; the German Parliament promptly voted to fund much of the cost of the enterprise.

Within a year, the German Deep-Sea Expedition was realized and the steamship *Valdivia* was chartered and fully outfitted with the latest marine technology, scientific facilities, and supplies. This included a research library stocked with the complete fifty-volume set of *Challenger* reports and a well-provisioned wine cellar, provided at cost by the patriotic directors of the Hamburg-American Company that owned and outfitted the *Valdivia*. In late July of 1898, the *Valdivia*, under the scientific leadership of Carl Chun and accompanied by an outstanding team of scientists, sailed from Hamburg's harbor and the expedition was under way.

Under Chun's leadership, the *Valdivia* sailed over 59,264 kilometers (c. 32,000 nautical miles) on a route that rounded the British Isles—stopping to visit the *Challenger* office in Edinburgh, where Chun and his colleagues examined some of the deep-sea organisms and deposits collected during that expedition. They then sailed south, skirting the African continent, past Prince Edward Island to the edge of the Antarctic ice sheet, then northward across the Indian Ocean as far as the island of Sumatra. The *Valdivia* returned by way of the Suez Canal and the Mediterranean, arriving back in Hamburg on April 28, 1899.

The expedition was a tremendous success and resulted in the discovery of numerous new species, some from depths greater than 4,000 meters (c. 13,000 feet), far deeper than had ever previously been sampled. Research on the specimens and oceanographic data amassed during the expedition occupied more than seventy scientists, and final editing of the twenty-four-volume series *Wissenschaftliche Ergebniss der Deutschen Tiefsee-Expedition auf dem Dampfer "Valdivia"* 1898–1899 was not completed until 1940.

Chun edited the series, and wrote the magnificent section on cephalopod mollusks, including the description of the notorious *Vampyroteuthis infernalis* (see page 160). His cephalopod specimens, like so many others described in the *Valdivia* series, were stunningly illustrated by the talented "scientific draughtsman and photographer" Fritz Winter, who had been a participant on the expedition and had had the opportunity to sketch many of his subjects while they were still alive. Chun also found time to write an illustrated, popular account of the journey, which was published under the title *Aus den Tiefen des Weltmeeres: Schilderungen von der Deutschen Tiefsee-Expedition*, which appeared in two volumes, the first published in 1900 and the second in 1903. These books were hugely popular and provided the German public with a wonderful insight into life aboard the *Valdivia*, the characters and personalities involved (accompanied by some beautifully executed caricatures), as well as the places visited and the scientific highlights of their discoveries.

2.

2. The cover of the first volume of Chun's popular account of the voyage of the *Valdivia* is beautifully decorated with marine motifs.

3. In his scientific account of cephalopods collected on the *Valdivia,* Chun described this impressive species of the deep-sea whiplash squid *Chiroteuthis imperator.*

4 & 5. This unusual octopus, *Velodona torgata*, depicted here in two views, was collected by Chun off the coast of east Africa at a depth of 749 meters (c. 2,500 feet). The extensive webbing, reaching to the tip of each arm, is highly characteristic and not found in other species.

After publication of *Aus den Tiefen des Weltmeeres,* Chun continued editing the scientific proceedings of the expedition until his death in 1914, at which time the editorial role was taken over by his protégé, August Brauer (see page 145). Quite remarkably, one man—Carl Chun—had instigated and realized one of his nation's most ambitious marine expeditions and firmly established Germany's place at the forefront of deep-sea oceanographic research.

Squid are by far the most numerous of the three main groups of coleoid cephalopods (octopuses, squid, and cuttlefishes). There are more than 450 species of squid living in the oceans, more than twice the number of octopus species. Teuthologists (those who study cephalopod mollusks) place most squid in the Order Teuthida; these are readily distinguished from octopuses by the presence of ten—rather than eight—arms around their mouths. The fourth and seventh arms of squid are modified into long, slender tentacles which are use to ensnare or grab their prey. Octopuses, on the other hand, have eight more or less equal-size arms, although in males, one of the arms is modified to transfer packets of sperm (spermatophores) into the female's body.

Most squid are relatively small, usually no more than 0.6 meters (c. 2 feet) long, but the colossal squid (*Mesonychoteuthis hamiltoni*) is estimated to reach lengths of 12–14 meters (c. 39–46 feet), making it likely the largest of all living invertebrates. Unlike the other contender for biggest invertebrate, the giant squid (*Architeuthis dux*), the colossal squid's tentacles are armed with a series of sharp hooks rather than suckers. The Giant Pacific octopus (*Enteroctopus dofleini*) is also a massive animal, reaching a record size of 9 meters (c. 30 feet) across, although, as with squid, most octopus species are much smaller than this. Predators of these large cephalopods are mainly whales, particularly sperm whales, although some sharks, elephant seals, and even sea otters and large fish have been found with remains of such cephalopods in their stomachs.

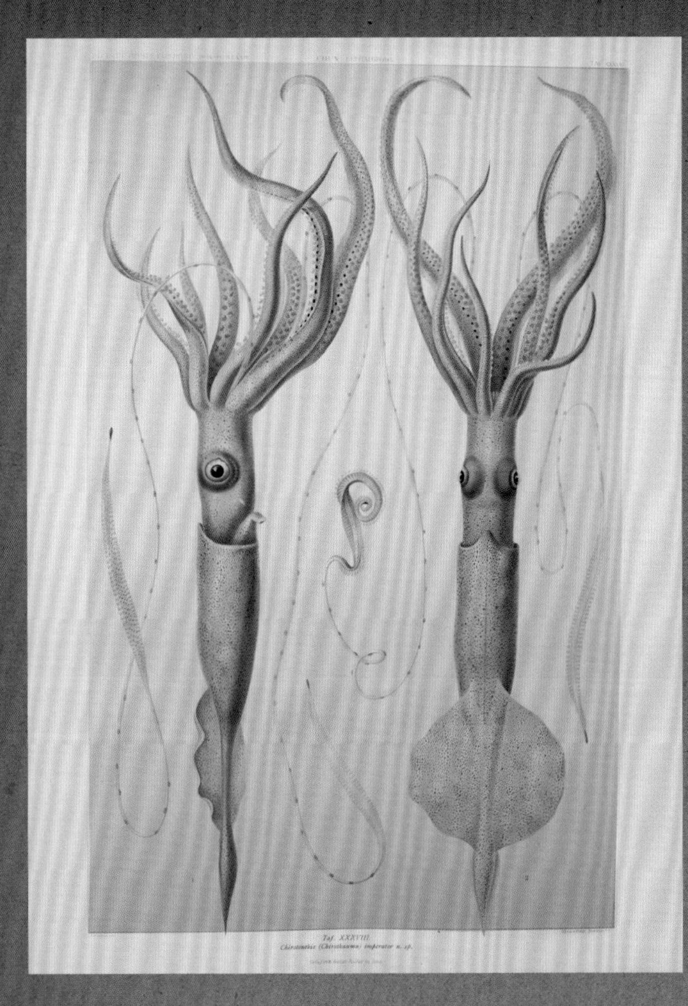
Taf. XXXVIII.
Chiroteuthis (Chirothauma) imperator n. sp.

3.

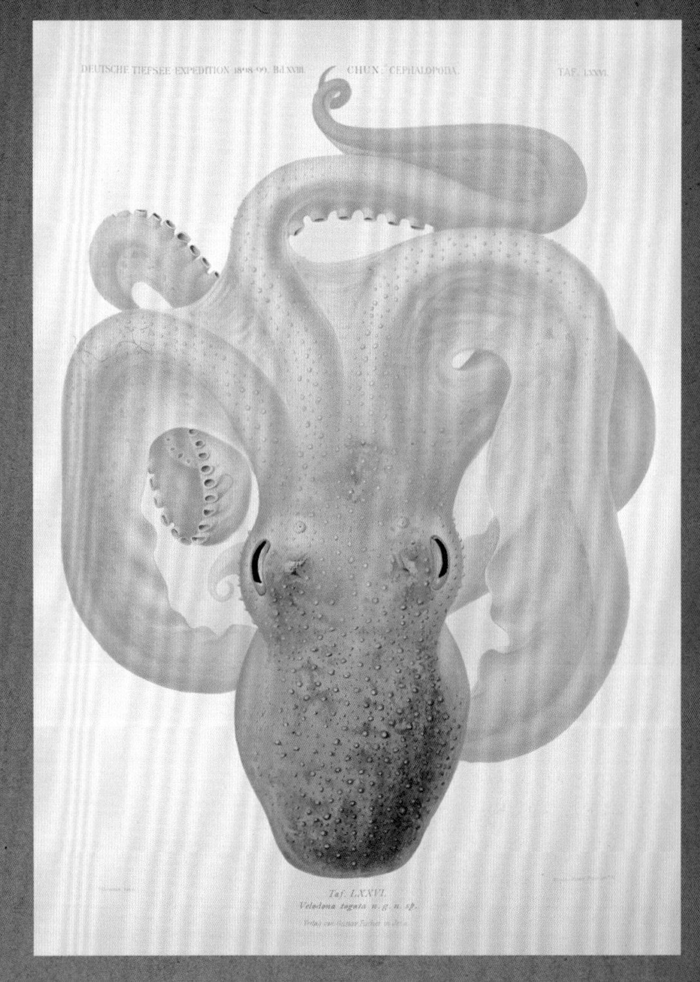
DEUTSCHE TIEFSEE EXPEDITION 1898-99. Bd.XVIII.
CHUN: CEPHALOPODA.
TAF. LXXVI.
Taf. LXXVI.
Velodona togata n. g. n. sp.

4.

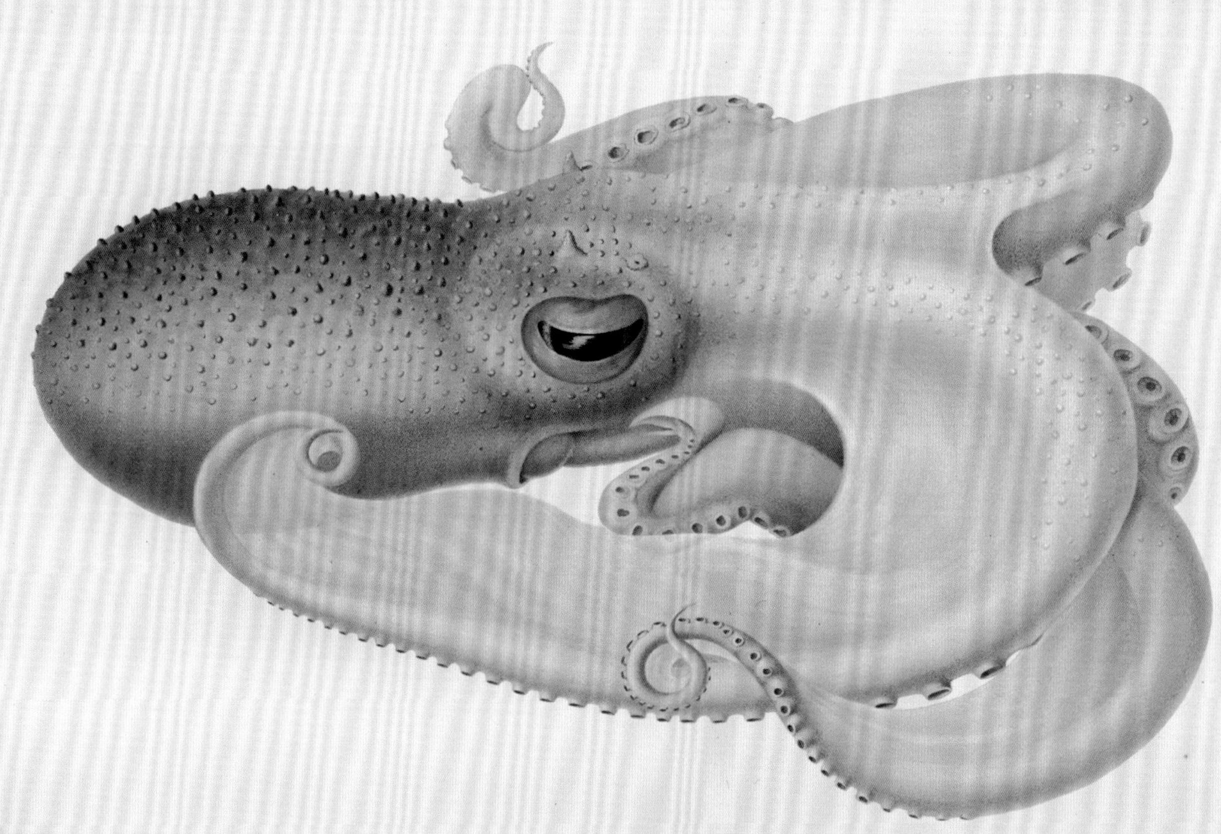
Taf. LXXV.

5.

12.

Scyphozoans: The Only *True* Jellies

Author
Ernst Vanhöffen (1858–1918)

Title
Die acraspeden Medusen der deutschen Tiefsee-Expedition, 1898–1899. (Wissenschaftliche Ergebnisse der Deutschen Tiefsee-Expedition auf dem Dampfer "Valdivia" 1898–1899, Band 3, Heft 1.)

(The acrasped medusae of the German Deep-Sea Expedition, 1898–1899. [Scientific results of the German Deep-Sea Expedition, on the steamer "Valdivia" 1898–1899, vol. 3, part 1.])

Imprint
Jena: G. Fischer, 1902

1. This beautiful Amakusa jellyfish (*Sanderia malayensis*) belongs to the "flag mouth" group, and its scalloped umbrella lacks the deep groove characteristic of the crown jellies.

The German biologist Ernst Vanhöffen was an accomplished field naturalist and an expert on the biology of jellyfish. Throughout his career, Vanhöffen participated in many major marine expeditions, and produced meticulously researched works on these complex soft-bodied marine invertebrates, many of which were most beautifully illustrated.

After graduating from the University of Königsberg in 1888 he left Germany to occupy the university's prestigious "table" at the Zoological Station in Naples, and continued his studies of jellyfish. While in Naples, Vanhöffen gained increasing international repute as a specialist on the group, and shortly after his return to Germany, the geophysicist and polar explorer Erich von Drygalski (1864–1949) selected him to join an expedition to western Greenland. In testimony to his growing expertise and flare for field biology, Vanhöffen served as von Drygalski's sole onboard naturalist during the Greenland voyage.

Returning to Germany in 1893, after an arduous winter in the Arctic, von Drygalski's expedition was deemed a national success and Vanhöffen was offered an appointment at the Zoological Institute at the University of Kiel (see page 127). Vanhöffen's activities during the Greenland expedition along with his knowledge of jellyfish had not escaped the notice of the eminent German biologist Carl Chun (see page 135), who requested his participation on the German Deep-Sea Expedition of 1898–1899. Thus it was that Vanhöffen joined Chun's select team of scientists and artists aboard the steamship *Valdivia* on its famous voyage of discovery in the deep seas.

Vanhöffen was assigned the task of sorting, classifying, and describing all species of jellyfish encountered on the epic voyage, and in 1902 he published the *Die acraspeden Medusen der deutchen Tiefsee-Expedition, 1898–1899*, one of two works on the jellyfish collected during the expedition. In this work, Vanhöffen provided detailed accounts of the strangely beautiful crown jellies, many of which were new to science. He named one of the animals illustrated in the accompanying plate *Periphyllopsis braueri*, in honor of his shipmate, the expedition's deep-sea fish expert, August Brauer (see page 155). Ew. H. Rübsaamen drew the stunning images accompanying Vanhöffen's treatise, and the lithography was that of Fritz Winter, the shipboard artist who worked closely with the scientists during the preparation of the twenty-three volumes published on the expedition's collections and oceanographic findings.

Early in 1901, Vanhöffen was appointed professor of zoology at Kiel, and later that year he joined his former colleague Erich von Drygalski as naturalist on the German South Polar Expedition of 1901–1903, aboard the newly built research vessel the *Gauss*. At the time, Antarctica was terra incognita, and despite international scientific collaborations—forcefully advocated by von Drygalski—political rivalries in the region were intense. The British had concurrently sent the British National Antarctic Expedition, led by Robert Falcon Scott (1868–1912), and the French and Swedish initiated similarly tasked exploratory voyages.

By February 1902, the *Gauss* became icebound some 50 miles off the Antarctic coast and remained beset for a full year. Fortunately, because the ice was not drifting, the German team was able to establish a fixed station from which to undertake a series of reconnoiters, and continue scientific measurements and collections. Finally, in February 1903, the ship came free and attempted to reach higher latitudes, but financial constraints forced von Drygalski to abandon the mission and sail home. Scientifically, the expedition had proven remarkably successful, but being trapped in the Antarctic ice was seen as a national failure that disappointed its sponsor, Emperor Wilhelm II, who saw only that Scott's expedition had been able to plant the British flag further south than had his German team.

Jellyfish of the Class Scyphozoa are close relatives of anemones (see page 92), corals (see pages 52 and 96), siphonophores (see page 44), and box jellies (see page 49), which together form the large Phylum Cnidaria. The name Scyphozoa is derived from the Greek word *skyphos*, which refers to a kind of drinking cup, obviously in reference to the familiar upturned bell or umbrella of the dominant life stage of this group. Most scyphozoans have a two-phase life history—as do other cnidarians—but in jellyfish, it is the medusa or jellyfish form that is dominant, while the sedentary hydroid or polyp phase is inconspicuous. Jellyfish, or jellies, are found in all oceans and they occupy habitats from the surface to the extreme depths. More than two hundred species have been scientifically described, but most experts agree that many more remain to be formally identified.

Scyphozoans, or true jellies, are classified into three main groups; the one to which Vanhöffen's belong is the Order Coronatae. These are commonly called crown jellyfish, and all have a deep groove that runs the circumference of their bell, or umbrella, giving them their distinctive crownlike shape. Most crown jellies, particularly in colder regions, are found only at great depth, although in the tropics some species live near the ocean surface. Some crown jellies, as well as many other cnidarians, are bioluminescent, meaning they can generate light in their tissues.

Cnidarians utilize a special kind of luciferin (the light-producing chemical) called colenterazine, a feature they share with bioluminescent ctenophores (comb jellies). Ctenophores (see page 49) and cnidarians used to be grouped together in the Phylum Coelenterata, but this relationship is no longer supported and it is thought that the two groups may have evolved colenterazine independently.

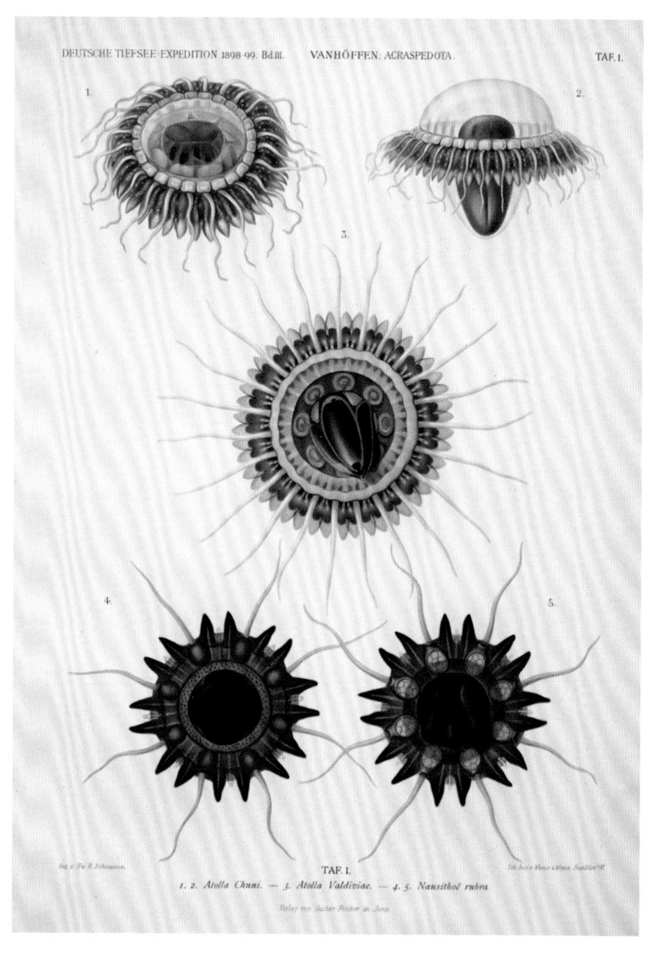

2.

2. Vanhoffen named the crown jellyfish (*Atolla chuni*) in honor of expedition leader Carl Chun and for their steamship *Valdivia* (*Atolla valdiviae*). Subsequent work has shown the latter is a synonym of *Atolla bairdii*, a species described some sixteen years earlier, but Chun's *Atolla* remains a valid name.

3. Based on this single individual (top right) Vanhoffen described a new genus of crown jelly (*Periphyllopsis*) and named the species in honor of his shipmate August Brauer. Since Vanhoffen's first discovery of *Periphyllopsis braueri*, only a few more individuals of this curious species have been found.

Gez. v. Ew. H. Rübsaamen. Lith. Anst. v. Werner & Winter, Frankfurt a/M.

TAF. II.

6 u. 8. *Periphylla regina.* — 7. *Periphyllopsis Braueri.* — 9. *Periphylla hyacinthina.*

Verlag von Gustav Fischer in Jena.

PTEROIS VOLITANS (LINNÆUS)

The Bold and Beautiful Lionfish

Author
David Starr Jordan
(1851–1931)

Title
The fishes of Samoa: description of the species found in the archipelago, with a provisional check-list of the fishes of Oceania. (Bulletin of the Bureau of Fisheries, v. 25, art. 5.)

Imprint
Washington: Government Printing Office, 1906

1. The spines of the red lionfish (*Pterois volitans*) are highly venomous, but are used entirely for defense. In hunting they use, instead, their large pectoral fins to corner small prey, which they then engulf whole.

David Starr Jordan was a towering figure in North American science of the late nineteenth and early twentieth centuries, and one of the best known naturalists and educators of his time. He was born in Upstate New York to progressive parents who shunned religious stricture and traditional social mores—characteristics that he inherited and which stayed with him throughout his life. Remarkable for the time, from the age of fourteen until graduation, Jordan attended a local girl's high school, and spent much of his time outdoors observing nature, collecting butterflies, and cataloging local plant life.

At the age of eighteen, he enrolled at Cornell University, where he majored in botany and excelled as both a student and a teacher. After graduation, he taught at various small Midwestern colleges and was extremely popular with his students, whom he took with him during summer breaks on collecting trips throughout the region and into the southern states.

Jordan was a lifelong proponent of the influential Louis Agassiz's (1807–1873) aphorism that biologists should "study nature, not books." Jordan had come under Agassiz's influence while visiting his newly established school for natural history and marine biology on Penikese Island off the Massachusetts coast, a school that was to become the inspiration for the world-renowned Marine Biological Laboratory established by Agassiz's students at Woods Hole in 1888. It was Agassiz who initially advised Jordan to take up ichthyology, and this was advice Jordan took to heart. In the years that followed, David Starr Jordan rapidly rose to prominence as one of the most influential of all American ichthyologists, the one to whom it is said that virtually all North American ichthyologists can trace back their professional ancestry.

In 1879 Jordan was appointed professor of zoology at Indiana University, and by 1884 his oratory skills, administrative acumen, and the popularity of his lectures and field trips made him an ideal choice for university leadership. He was appointed university president at the age of thirty-four and immediately came to prominence as the nation's youngest. His presidency at Indiana was highly successful and under his tenure the institution rose in academic stature and financial stability. Despite heavy administrative demands, Jordan remained scientifically active and it was said that scarcely a year of his presidency passed when his publication output did not exceed that of many members of the Indiana science faculty.

Jordan's successes did not escape notice and his reputation as a progressive educator and persuasive proponent of secular coeducation attracted the interest of the former California senator Leland Stanford and his wife, Jane. The couple was planning to build an academic institution in the American West to match the prestigious schools of the East Coast. But as a contrast to many of those schools, the Stanfords wanted to build a university and a museum that was coeducational, nondenominational, and, above all, practically geared to produce "cultured and useful citizens." Jordan was offered the job of president, and at the age of forty, he left Indiana for Leland Stanford's Palo Alto Stock Farm and began the foundational work that was to become Stanford University—an institution he guided through tumultuous early years and helped to establish as a center of academic excellence. From the opening of its doors in 1891 until he stood down in 1913, Jordan served as president of Leland Stanford Junior University, so named in memory of the Stanford's only child, who had died of typhoid fever at the age of sixteen, but more commonly known today as Stanford University.

Early in Jordan's career, he had come to the attention of Spencer Fullerton Baird (1823–1887), the influential secretary of the Smithsonian Institution in Washington, D.C., and the first commissioner of the newly established United States Commission of Fish and Fisheries. The commission had been created to assess the nation's fish and marine resources, determine whether their status was in decline, and, if so, make recommendations for remedial action. Baird provided Jordan with financial support and access to government facilities, and Jordan worked closely with the Smithsonian and the Fish Commission (which in 1903 was reorganized into the U.S. Bureau of Fisheries, the precursor of today's National Marine Fisheries Service) for most of his career. He published many studies in the scientific series of the commission; the work *The fishes of Samoa: description of the species found in the archipelago, with a provisional check-list of the fishes of Oceania* appeared in the 1905 volume of the U.S. Bureau of Fisheries bulletin series.

Jordan's expedition to what was then American Samoa, like so many others that he led, was made under the auspices of the bureau and contributed greatly to knowledge of the diversity and distribution of fishes across the United States and its overseas territories.

During the 1902 Samoa expedition, Jordan wrote that "the coral reefs of the South Seas literally swarm with fishes," and, quoting from a record of the voyage of Captain Cook, he noted that "their colors were the most beautiful that can be imagined: blue, yellow, black, red, etc., far excelling anything that can be produced by art."

Many of the colored plates that accompany his treatise were based on field sketches drawn by Jordan himself during the expedition. Modestly, he acknowledged that some of those sketches were "not entirely accurate in certain details of scales and fin rays, but the shades of coloration are very well shown."

One of the most striking of these images is of a lionfish of the genus *Pterois*, a name derived from the Greek word *pteron*, which means wing or fin, and refers to the very large winglike pectoral fins that are so characteristic of these fishes. There are ten species of lionfish in the genus *Pterois*, and all are known for their bold patterning and beautiful coloration as well as for their venomous dorsal fin spines.

Such striking coloration in a venomous animal is considered to be an example of aposematism—a warning coloration or signaling that serves as an antipredator device by advertising to potential predators the inadvisability of attack, which in the case of the lionfish could result in envenomation of the attacker. Such a system is of obvious benefit to both the predator and prey, as both can potentially avoid harm. As a result, there are few natural predators of adult lionfish, although juvenile lionfish are vulnerable to cannibalism by adults and are potential prey of numerous other larger fish species.

2. The eye of the sea tilefish *Malacanthus latovittatus* is a relatively common Indo-Pacific reef species. It, like many tilefish, is known to have particularly acute vision, which is used in hunting.

3. Jordan, like Captain James Cook before him, was struck by the vibrant colors of many of the reef fishes he encountered. Here, he illustrates three beautiful tropical wrasses of the genus *Halichoeres*.

4. Both of these resplendent reef fishes, the Chevron butterflyfish (*Chaetodon trifascialis*) (above) and the orangespotted filefish (*Oxymonacanthus longirostris*) (below) feed exclusively on coral polyps.

OCEANOPS LATOVITTATA (LACÉPÈDE)

2.

1 HALICHŒRES TRIMACULATUS (QUOY & GAIMARD)
2 HALICHŒRES DÆDALMA JORDAN & SEALE
3 HALICHŒRES OPERCULARIS (GÜNTHER)

3.

1 MEGAPROTODON TRIFASCIALIS (QUOY & GAIMARD)
2 OXYMONACANTHUS LONGIROSTRIS (BLOCH & SCHNEIDER)

4.

18. TRIOPHA MACULATA MACFARLAND
Dorso-lateral view, about 1.5 times natural size

MacFarland's Sea Slugs

Author
Frank Mace MacFarland
(1869–1951)

Title
Opistobranchiate Mollusca from Monterey Bay, California and vicinity. (Bulletin of the Bureau of Fisheries, v. 25, art. 3.)

Imprint
Washington: Government Printing Office, 1906

Figure 2
Author: Salvatore Trinchese (1836–1897)
Title: *Æolididae e famiglie affini del porto di Genova*
Imprint: Bologna: Tipi Gamberini e Parmeggiani, 1877–1881

1. The speckled triopha (*Triopha maculata*) is a large nudibranch sea slug, which can grow to 18 centimeters (c. 7 inches) long. MacFarland first discovered this species living in tide pools of Monterey Bay.

Frank Mace MacFarland spent a long and productive career devoted to the study of nudibranch sea slugs, some of which are the most beautifully colored and ecologically complex of all invertebrate inhabitants of in-shore marine habitats. His studies, particularly of the species-rich communities of California, were considered to be among the most detailed and meticulous of the time, and at his death in 1951, he was broadly acknowledged to be a world expert on this ecologically diverse group of mollusks.

MacFarland was born in Centralia, Illinois, and studied at DePauw University in Indiana, graduating in 1889. For three years, he taught biology and geology at Olivet College in Michigan. In 1892 he left Michigan to enroll as an advanced student and instructor at the newly established Leland Stanford Junior University in Palo Alto, California. As a student, he focused his researches on the sea slugs of Monterey Bay and their allies. The work *Opistobranchiate Mollusca from Monterey Bay, California and vicinity* was begun while he was based at the newly constructed Hopkins Seaside Laboratory—the marine research facility that had been founded by David Starr Jordan after he was appointed president of Stanford University (see page 143). In 1893 MacFarland was awarded a master's degree and, although he was to maintain an association with the university throughout his career, he decided to carry out the remainder of his graduate studies in Europe rather than at Stanford. In the following years, he studied at the universities of Zürich and Würzburg and in the early 1900s he spent a year working at the Zoological Station in Naples.

While at the Zoological Station, MacFarland was able to make detailed studies of nudibranchs and their allies, and to compare the Mediterranean species with those he had studied along the California coast. On his return to the United States, MacFarland was appointed professor of histology (the study of the microscopic structure of cells and tissues) at Stanford, where he continued to work on his sea slugs and their anatomical structures for the rest of his career.

Throughout his career, MacFarland's wife, Olive Hornbrook MacFarland, assisted him as an illustrator. Many beautiful representations of nudibranchs, often drawn from life, were prepared by her to accompany her husband's publications. A particularly attractive example of her work accompanys this essay—a beautiful illustration of a living specimen of speckled triopha (*Triopha maculata*), a species discovered and described by MacFarland from tide pools of Monterey Bay.

Opisthobranchians (sea slugs, bubble shells, and sea hares) are a highly specialized group of mollusks, and among the most colorful and ecologically diverse of them are the three thousand or so species of nudibranch (naked gilled) sea slugs. These soft-bodied marine mollusks have lost all trace of an external shell, are bilaterally symmetrical, and bear a prominent pair of tentacles close to the mouth, which are used for orientation. Behind the tentacles are distinctive, club-shaped olfactory organs called rhinophores. In addition to tentacles and rhinophores, many nudibranchs have a series of fleshy outgrowths (parapods) along the side of their muscular "foot," and highly elaborate and often colorful gills arrayed over their bodies or clustered around their posterior ends.

Because nudibranchs lack a protective external shell, they are vulnerable to predation and as a result have evolved some of the most spectacular defense mechanisms known in the animal kingdom. Some species are able to synthesize toxic compounds, including such nocuous ones as sulphuric acid—which they produce as a repellent to predators. Others store compounds taken from their prey (often from toxic algae) to deter predators. And perhaps the most remarkable of all such antipredator devices is found in many of the nudibranchs that feed on cnidarians (jellyfish, anemones, corals, and hydrozoans). Using some unknown mechanism, these sea slugs are able to "hijack" the stinging cells (nematocysts; see page 49) of their prey and repurpose them as a defensive arsenal of their own.

Perhaps it is not surprising then that many nudibranchs are strikingly patterned and colored—clearly advertising to potential predators the inadvisability of attack.

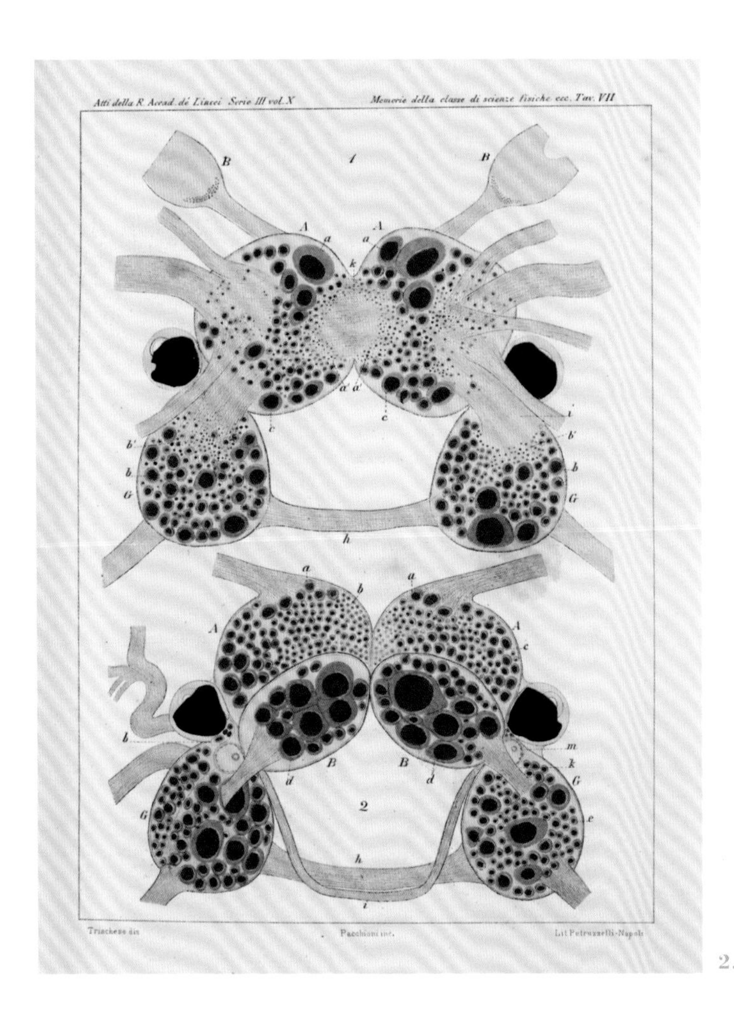

2.

2. The Italian malacologist Salvatore Trinchese wrote an early work on Neapolitan sea slugs, and his studies served as a foundation for MacFarland's own. This plate, illustrating the complex nudibranch neural anatomy, was published by Trinchese in 1881.

3. Viewed from above, this grand triopha (*Triopha occidentalis*) displays a characteristic rosette of branchial plumes (gills) located on its posterodorsal surface.

4. *Felimare porterae* is another of the beautifully colored nudibranch sea slugs collected by MacFarland and drawn by Olive Hornbrook MacFarland.

19. TRIOPHA GRANDIS MACFARLAND
Dorsal view, slightly larger than life

A. HOEN & CO., LITH.

3.

BULL. U. S. B. F. 1905

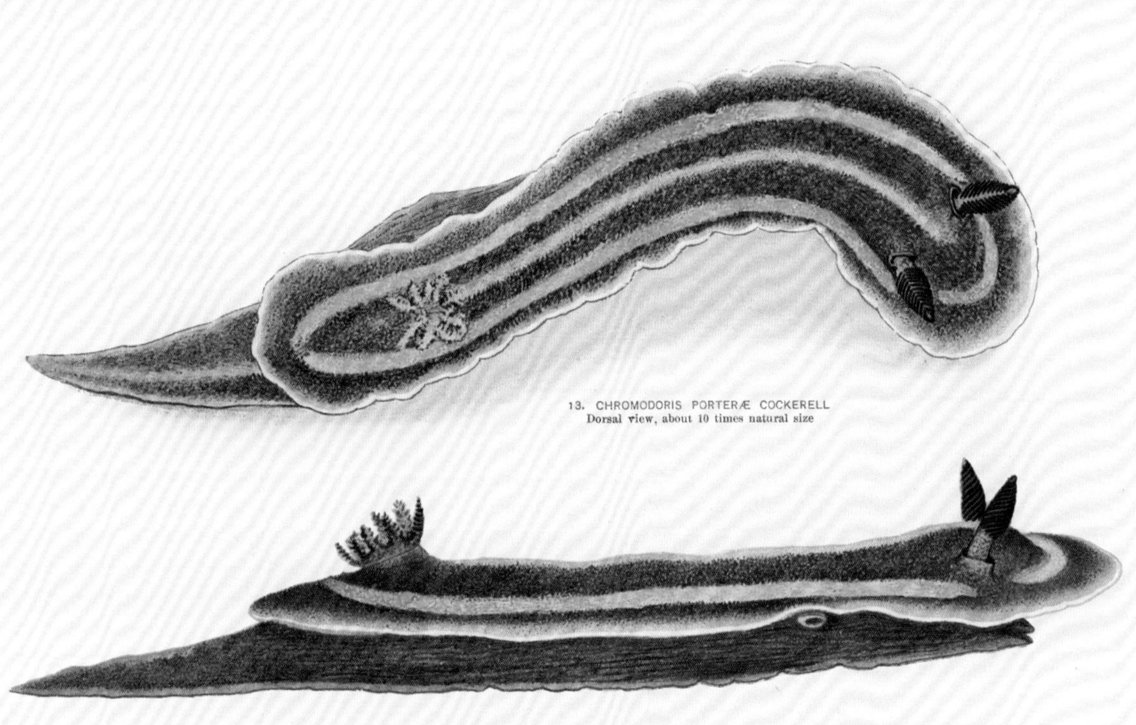

13. CHROMODORIS PORTERÆ COCKERELL
Dorsal view, about 10 times natural size

14. CHROMODORIS PORTERÆ COCKERELL
Lateral view, about 10 times natural size

A. HOEN & CO., LITH.

PLATE XXVI

4.

4/5

Petrels: Bird Life at Sea

Author
Frederick DuCane Godman
(1834–1919)

Title
A monograph of the petrels (order Tubinares)

Imprint
London: Witherby & Co.,
1907–1910

Frederick DuCane Godman was in many ways the archetypal "gentleman naturalist" that was so characteristic of the age and of his social class. Born to a wealthy family, at the age of ten he was sent to attend Eton College, an elite English boarding school near Windsor. Ill health caused his early return to the family home in Park Hatch, Surrey, where he was educated by a series of private tutors. Despite a frail constitution, Godman spent much of his time outdoors occupied in the leisured pursuits of his class—hunting, shooting, and fishing on the grounds of the family estate.

While at Park Hatch, he began to study local insects, particularly butterflies and moths, as well as birds and the region's flora. In 1853 he attended Trinity College, Cambridge, and there became acquainted with other young men who shared his enthusiasm for natural history. His friendship with one in particular, the ornithologist Osbert Salvin (1835–1898), began a lifelong collaboration. At Cambridge, Godman and Salvin met regularly with others interested in birds and hatched the idea of forming an ornithological union to encourage their study and conservation. In 1858 the British Ornithologists' Union (BOU) was founded and its prestigious quarterly journal, *Ibis*, began publication the following year.

Godman's large fortune allowed him to travel the world and pursue his scientific interests with no regard to paid employment. In 1861 he joined Salvin, who was traveling in Central America, and the two spent time exploring Guatemala, Jamaica, and Belize together with Salvin's wife. Godman's health failed during the trip and he was forced to leave his friends to return to England. By 1876 Salvin and Godman embarked on the ambitious project of writing a flora and fauna of Central America. They assembled enormous numbers of specimens and, with Godman's fortune, purchased many additional collections. Most of the huge Salvin-Godman collection, including a staggering eighty thousand bird skins, was ultimately given to the British Museum of Natural History and, between 1879 and 1915, a total of sixty-three volumes comprising the monumental *Biologia Centrali-Americana* were published.

The death of Salvin in 1898 was a tremendous blow to Godman, who had felt of Savin "more intimately connected than most brothers," and left him with much of his friend's unfinished works to complete. Beyond completion of the *Biologia Centrali-Americana*, Godman turned his attentions to what was to become *A monograph of the Petrels (order Tubinares)*. In the preface to that work, Godman explained that he and Salvin had amassed large collections of petrels in anticipation of producing a monograph on these poorly studied birds. The famous Dutch illustrator Johannes Gerardus Keulemans (1842–1912) had been commissioned to produce a series of lithographic plates for the work, and Godman determined to complete his friend's "long intended Monograph." He acknowledged that he was no expert on this taxonomically difficult

1. It is likely that the artist Johannes Keulemans saw only dried skins of the birds he illustrated for Godman. It is testament to his talent that his drawings, like this elegant fork-tailed storm petrel (*Oceanodroma furcata*), appear so alive and naturally posed.

group of birds, and were it not for the promised assistance of Richard Bowdler Sharpe (1847–1909), then head of the British museum's bird collection, "it would have been presumptuous in me to have undertaken so difficult a task." But with Sharpe's assistance, Godman completed the monograph, and it was to Sharpe's two daughters that he entrusted the work of accurately hand-coloring Keulemans's beautiful lithographic drawings. The monograph appeared in two volumes, the first in 1907 and the second in 1910. Sadly, Sharpe did not live to see the publication of the second volume as he died of pneumonia in the winter of 1909, just a few weeks before it appeared.

Godman died peacefully at his home in February of 1919. A pillar of the Victorian establishment, Godman had been elected fellow of the Royal Society in 1882, served as trustee of the British museum, and received the coveted Gold Medal of the Linnean Society. From 1896 to 1913, he served as the third president of the British Ornithologists' Union, and on his death the union instituted the Godman-Salvin Gold Medal—its highest honor for outstanding work in ornithology. This posthumous honor, which forever links his name to that of his dear friend, likely would have given Godman the most satisfaction of all.

Petrels are the consummate marine birds—they spend their entire lives at sea, coming to land only to breed. They are currently classified in the Order Procellariiformes (formerly known as Tubinares) and are commonly called tubenoses for the curious anatomy of their nostrils, which are enclosed within conspicuous bony tubes. Among tubenoses, petrels are divided into two main groups: storm petrels and shearwaters. A third group of tube-nosed procellariiforms are albatrosses, which although not considered petrels, are closely related to them.

Storm petrels, like the elegant fork-tailed petrel (*Oceanodroma furcata*) so beautifully rendered in the accompanying plate, are the smallest seabirds, ranging from just 12 to 25 centimeters (c. 5 to 10 inches) in length. Storm petrels are reputedly named for their habit of flying in the lee of ships during bad weather. Despite their small size, storm petrels roam the open oceans, descending only to feed on the wing by hovering and snatching small fish and crustacean larvae from the water surface. Petrels are so marine adapted that their legs, although quite long, are incapable of supporting their bodies for more than a few steps on land.

Shearwaters are larger and display a characteristically stiff-winged shearing flight that requires a minimum of energy expenditure. Many are spectacular long-distance migrants with some, such as the sooty shearwater (*Puffinus griseus*), regularly traveling well over 500 kilometers (c. 300 miles) every day. Unlike storm petrels, shearwaters are diving hunters and feed on fish and squid taken from as deep as 60 meters (c. 200 feet) or more, although they will also take food from the water surface and often follow fishing boats or whales in the hope of scooping up food scraps. Shearwaters are very long-lived, with some species commonly surviving to well beyond fifty-five years of age.

Albatrosses are the largest of all flying birds and some, such as the Royal (*Diomedea regia*), have wings spans reaching 3.7 meters (c. 12 feet). Twenty-one species are recognized, with most ranging the southern hemisphere. Like shearwaters, albatrosses are long-distance migrants and range the oceans descending only to feed, mainly on squid, fish, and crustaceans, which they grab on the wing from the water surface. Albatrosses almost always return to their natal sites to breed, and this tendency is so strong that in some species the average distance between hatching sites and the sites where the birds establish breeding territories can be only 21 meters (c. 70 feet).

2. The European storm petrel (*Hydrobates pelagicus*), a small, square-tailed bird with a curious batlike flight pattern, was once considered a bad omen by North Sea sailors because of its association with turbulent weather.

3. Sooty Shearwaters (*Puffinus griseus*), named for their dark plumage, congregate in vast numbers on nesting colonies. Females lay a single egg inside long plant-lined burrows, where the chick remains protected until fledged. Known as muttonbirds in New Zealand, young chicks are traditionally harvested by Maori for oil and meat.

4. *Thalassarche salvini* (Salvin's albatross), considered a "mollymauk," or smaller Albatross, has a wingspan of over 2.5 meters (c. 8.5 feet). Named in honor of Osbert Salvin, this beautiful bird ranges across the southern ocean landing only to breed on isolated rocky islands.

2.

3.

4.

1
3/1

Deeply Studied Deep-Sea Fish

Author
August Brauer (1863–1917)

Title
Die Tiefsee-Fische. (Wissenschaftliche Ergebnisse der Deutschen Tiefsee-Expedition auf dem Dampfer "Valdivia" 1898–1899, Bd. 15.)

(The deep-sea fishes. [Scientific results of the German Deep-Sea Expedition, on the steamer "Valdivia" 1898–1899, vol. 15.])

Imprint
Jena: Gustav Fischer, 1908

1. Despite their ferocious appearance, most deep-sea fishes are small animals, and this female black seadevil (*Melanocetus johnstonii*) grows no larger than 20 centimeters (c. 8 inches).

The German Deep-Sea Expedition of 1898–1899 under the leadership of Carl Friedrich Chun (see page 135) established his country's place at the forefront of the rapidly developing science of deep-sea oceanography. Chun assembled an outstanding team of scientists aboard the research vessel *Valdivia*, whose mission was to sample, record, and study as many deep-sea organisms and habitats as possible, throwing a much-needed light on that vast but poorly known ocean realm. The expedition was a resounding success, with countless new discoveries carefully documented in the monumental twenty-four-volume series that resulted. Of these volumes, none was more impactful than that of August Brauer, whose extraordinary report on *Die Tiefsee-Fische* appeared in 1908 as volume fifteen of the *Valdivia* series. Brauer's seminal work on deep-sea fishes, many encountered for the first time during the voyage of the *Valdivia*, is widely acknowledged as the starting point of deep-sea ichthyology and remains an invaluable reference to this day.

Brauer was born the youngest of nine children in the town of Oldenburg in Lower Saxony. His father was a successful merchant, but already as a child Brauer was captivated by the natural sciences and, at the age of nineteen, he left Oldenburg to study at the universities of Freiburg, Berlin, and Bonn. He completed his studies in Bonn, graduating in 1885 with a thesis on microscopic heterotrichs (ciliated unicellular eukaryotes). After military service, he taught briefly in a high school before being accepted in 1890 as an assistant at the Zoological Institute in Berlin. After a brief visit to Trieste to study brine shrimp development, he was appointed lecturer at the University of Marburg.

Brauer's zoological interests ranged widely, and in 1894 he spent eight months on the remote Indian Ocean island of Mahé in the Seychelles collecting and studying amphibians and reptiles. In 1898 Chun invited Brauer to join the scientific staff of the *Valdivia* and charged him with the daunting task of documenting the many deep-sea fishes collected. It is unclear what experience, if any, Brauer had with ichthyological studies prior to his momentous work on the *Valdivia*, but he rose to the task that Chun had so presciently set him. Brauer collaborated closely with the shipboard artist and scientific draftsman Fritz Winter, a friend whom he considered not just an outstanding artist but also a talented zoological observer.

Brauer fully acknowledged the tremendous importance of lifelike and accurate representations of the many extraordinary fishes he was discovering. The work was laborious and often frustrating as it soon became clear that catching fishes in deep waters was extremely difficult. The conditions encountered and the deployment of the complicated dredging equipment proved ineffective at catching fish from the deep ocean floor. Brauer was particularly disappointed by his inability to collect fishes in the Antarctic, where the dredging was done at depths of 4,800 to 6,100 meters (c. 16,000 to 20,000 feet), some of the deepest waters encountered on the voyage. Happily, he was more successful in gathering an extraordinary array of often bizarre-looking specimens from the then newly recognized bathypelagic zone—a huge, virtually lightless region of the ocean, from depths of 1,000 to 4,000 meters (c. 3,300 to 13,000 feet). This massive ocean biome had been recognized by Chun as being particularly rich in planktonic and cephalopod life and Brauer was able to confirm that the same was true for fishes.

In 1899 Brauer returned to Marburg and continued work assembling two magnificent sections, one on systematics and one on anatomy, for his report on *Die Tiefsee-Fische*. When the volume was finally published in 1908, Brauer's reputation was established, and at Chun's recommendation, he was appointed director of the Natural History Museum in Berlin. He proved an effective manager and fund-raiser, and under his leadership the museum expanded its holdings to include most of the zoological collections from the German Deep-Sea Expedition, as well as from the many German expeditions that were undertaken in this period.

Although mainly occupied with administrative duties, Brauer continued work on his deep-sea fishes. He also actively published on subjects as diverse as the relationships of hyraxes and a ninteen-part series on the freshwater fishes of Germany. After Chun's death in 1914, he assumed editorship of the *Wissenschaftliche Ergebnisse der Deutschen Tiefsee-Expedition auf dem Dampfer "Valdivia"* 1898–1899.

In September 1917, after returning from a holiday with his sisters, Brauer experienced chest pains. His doctor prescribed bed rest, but, sadly, just a few days later, he was found dead in his study with a book he had been reading still in his hands. His sudden passing, at just fifty-four years of age, was much lamented by colleagues and staff of the Berlin museum. In a heartfelt obituary, his friend and former shipmate from his days on the *Valdivia*, Ernst Vanhöffen (see page 139), wrote of his devotion to the museum, noting that before his death, Brauer had bequeathed his extensive library and not inconsiderable personal fortune to the museum he had so loved.

Until around the middle of the ninteenth century, extremely little was known of life below the sunlit ocean waters, and it was generally believed that in the absence of light and at such great pressures none could exist much below 600 meters (c. 2,000 feet). This long-standing abyssal or azoic theory was increasingly challenged as scientists began to sample in deeper and deeper waters, and by 1861, definitive proof of life at depth was established when a telegraph cable taken for repair from the Mediterranean floor at 2,000 meters (c. 6,600 feet) was found to be encrusted with living mollusks and corals. Once the existence of life at depth was established, the next great task was to establish its nature and distribution throughout the oceans. It was here that Brauer made his greatest contributions by firmly establishing the existence and widespread global occurrence of a rich bathypelagic fish fauna. His detailed anatomical studies further revealed an extraordinary array of sensory adaptations to low or absent levels of ambient light, and of numerous modes of biological light production through bioluminescence.

2. The head of a spectacular specimen of *Melanostomias melanops* (Family Stomiidae) (top right) shows details of its bioluminescent chin barbel and large "flashlight" photophore situated behind the eye. Living in the darkness of the deep sea, barbeled dragonfishes possess numerous bioluminescent structures.

3. In this plate Brauer depicts fishes from two distantly related families, the telescopefish, *Gigantura chuni* (Giganturidae), and the barreleye, or spookfishes, *Opisthoproctus soleatus* and *Winteria telescopa* (Opisthoproctidae). These fishes have independently evolved highly modified, tubular eyes that concentrate the faintest ambient, or bioluminescent, light.

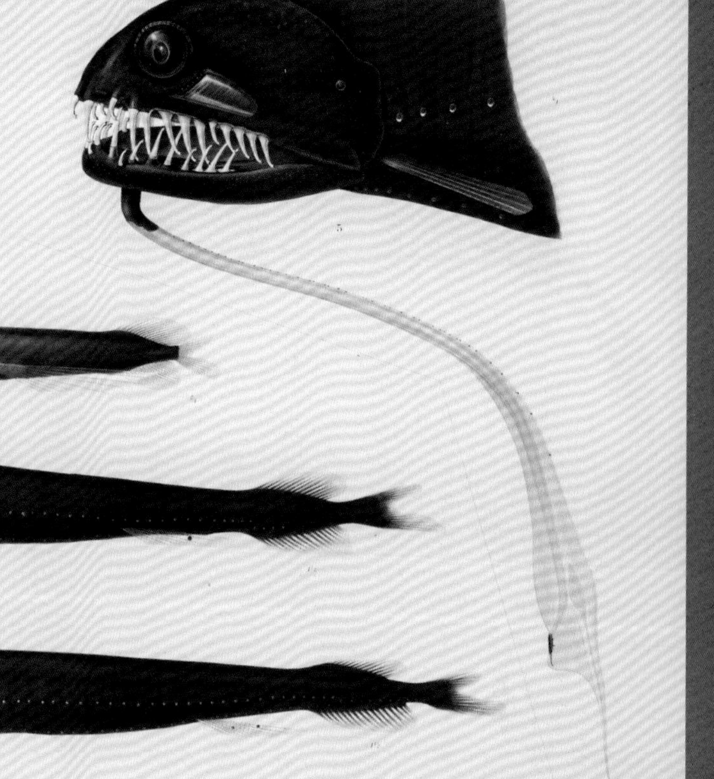

2.

3.

The Vampire Squid from Hell

Author
Louis Joubin (1861–1935)

Title
Résultats des Campagnes Scientifiques accomplies sur son yacht par Albert 1er Prince Souverain de Monaco. Fascicule LIV Céphalopodes provenant des Campagnes de la Princesse-Alice (1898–1910).

(Results of the scientific expeditions of Albert 1st, Sovereign Prince of Monaco on his yacht. Issue LIV Cephalopods from the expeditions of the Princess Alice [1898–1910].)

Imprint
Monaco: Imprimerie de Monaco, 1920

1. Despite its strange appearance and ominous name, *Vampyroteuthis infernalis* (literally the "vampire squid from hell"), this docile deep-water mollusk is the only known octopod that is not an active predator.

The French biologist Louis Marie Adolphe Olivier Edouard Joubin published many works on marine invertebrates but is probably best known for his studies of cephalopods (nautiloids, squid, octopuses, and cuttlefish). In the early 1880s, Joubin served as director of two of his nation's famous marine research stations at Roscoff and Banyuls-sur-Mer (see page 96), and later as an instructor at the University of Rennes.

A leading scientist of his day, Joubin was chosen as president of the prestigious French Zoological Society, and in 1906 he was appointed as chair of mollusks, worms, and zoophytes at the Natural History Museum in Paris. That same year, Joubin was selected by Albert I, Prince of Monaco, to oversee zoological instruction at the newly founded Oceanographic Institute that the prince had established in Paris.

Prince Albert I of Monaco (1848–1922), perhaps more than any other figure of the nineteenth century, helped to establish the nascent field of oceanography as a major scientific discipline. With a considerable fortune at his disposal, Albert commissioned a series of increasingly sophisticated and scientifically well-equipped yachts and in these, accompanied by leading marine biologists of the day, he led many expeditions, collecting, mapping, and deep-sea dredging around the globe. In 1899 he laid the foundation stone for what was to become the famous Oceanographic Museum of Monaco, which included a spectacular public aquarium and library, majestically situated on a rocky bluff overlooking the Mediterranean.

Joubin's scientific studies were greatly facilitated by the generous sponsorship of the prince, and in the introduction to the present work Joubin expressed his gratitude to "His Serene Highness Prince Albert 1st of Monaco, Member of the Institute of France, for the trust he has bestowed on me for a quarter of a century." The work, published in 1920, was the last in a series of papers written by Joubin on cephalopod mollusks collected between 1898 and 1910 during cruises of the *Princesse-Alice*, one of the prince's yachts. In these works, Joubin described a large number of highly modified deep-sea octopods, many of which he believed were new to science. One "singular cephalopod" had attracted his particular interest—it was a strange finned octopod, fully black in color, with large red eyes and two prominent, luminous organs (photophores) glowing on its back near the paired fins.

In 1912 Joubin had formally described the strange creature, giving it the name *Melanoteuthis lucens*, the shining black squid. However, even in 1920 when he reported again on this curious animal, it was apparent that Joubin was unaware of the work of his German contemporary, Carl Chun (1852–1914; see page 135). In 1903 Chun had published the description of a very similar deep-sea octopod that he had named *Vampyroteuthis infernalis*, the vampire squid from hell. As more specimens of this strange cephalopod became available for study, it was shown that Joubin's shining black squid was in fact none other than the vampire squid.

Because Chun's 1903 description of the animal preceded that of Joubin, the name *Vampyroteuthis infernalis* takes priority, and Joubin's *Melanoteuthis lucens* is today considered to be a synonym. So Joubin's shining black squid, beautifully rendered by the expedition's artist Miss Vesque and based on a watercolor of the live animal at the time of its capture, is actually an illustration of *Vampyroteuthis infernalis.*

Living squid, octopuses, and the vampire squid all belong to a group of cephalopod mollusks called coleoids. These are easily recognized by the absence of an external shell and the presence of armlike structures around their beaked mouths. Although there is still some question as to how the different groups of coleoid cephalopods are related to one another, the vampire squid, although commonly called a squid, is probably more closely related to octopuses than to true squid and cuttlefishes.

Worldwide there are over two hundred species of octopuses, but only one known species of vampire squid. Vampires have been found throughout temperate and tropical oceans, living mostly at great depths, between 600 and 900 meters (c. 2,000 and 3,000 feet). Ironically, we now know that not only are they not true squid, they are also not vampires—in fact, *Vampyroteuthis infernalis* is the only known cephalopod that is not an active predator. Instead it feeds on detritus (the remains and waste of numerous marine organisms) that rains down from the surface waters. The vampires collect this so-called "marine snow" (see page 128) in mucous produced by a modified pair of arms (velar filaments); it is then formed into balls of food that are passed to the mouth and ingested.

2. In this plate Joubin illustrated details of the anatomy of the luminous organs of a specimen of *Vampyroteuthis* that he dissected.

3. This beautifully colored whiplash squid (*Mastigoteuthis magna*) was first described by Joubin based on specimens trawled from deep water during the voyages of the *Princesse-Alice*.

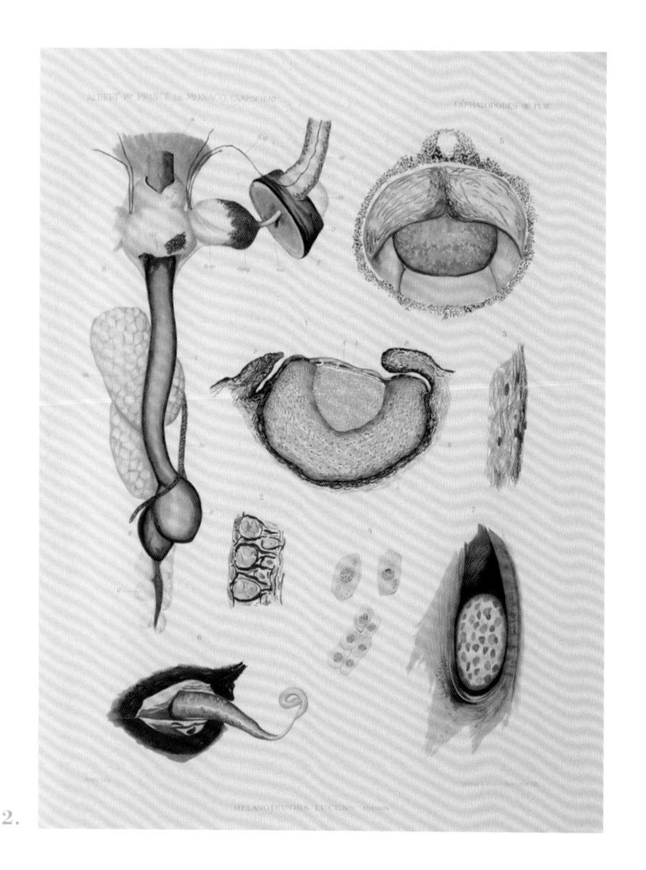

2.

3.

ACKNOWLEDGMENTS

Writing this book has been both a pleasure and an education, and for this there are a great many people to thank. First and foremost, the wonderful library staff that so patiently, and so often, made the journey with me into the vaults of the Museum's spectacular Rare Book Collection, not just to let me inside, but also to help in the search for the right volumes. Annette Springer, Gregory Raml, and Diana Shih unstintingly gave of their time, and Barbara Rhodes not only taught me how to handle those precious tomes but also patiently guarded over them during many hours of photography. I am indebted also to Barbara for contributing her insightful essay explaining the importance and complexities of book conservation.

In this paean to the wonderfulness of librarians, I want to express my profound gratitude to Mai Qaraman Reitmeyer, the epitome of helpful professionalism. Mai's extensive knowledge, so enthusiastically given, has made my task so much lighter. And, of course, to Tom Baione, the Library's director, not just for freeing up his staff to deal with my every request but also for his sound guidance and encouragement throughout the process.

Pictures are at the heart of this volume, and I am greatly indebted to Roderick Mickens for his outstanding photographic work. Rod, who shot every image, was a real pleasure to work with, and, characteristically, was always in good spirits despite many long hours of peering through a camera lens under hot lights.

At the American Museum of Natural History, my thanks go also to Alex Navissi, Will Lach, and Sharon Stulberg for their work on the series, and for keeping me on track and on time. And at Sterling Signature I am very grateful to my editor, John Foster. John has been a great advocate for scientific accuracy along with beautiful aesthetics, and I greatly appreciate his guiding me through this process. Also at Sterling, my thanks to Chris Thompson for art direction, Yeon Kim for design, Betsy Beier for editorial direction, and Sal Destro for production.

There are also many friends and colleagues who have helped in innumerable ways, from translation to scientific interpretation, and I would like to acknowledge Lita Elvers, Christine Matthei, John Maisey, Paul Sweet, and Saskia Grotenhuis for their input and advice.

Finally, last but certainly not least, to my wife Jackie Black, whose helpful reviews, sage advice, and tireless support allowed me to make all of the deadlines while keeping me smiling.

Thank you, all.

SUGGESTED READING

Cramer, Deborah (2001). *Great Waters: An Atlantic Passage*. New York, NY: W. W. Norton & Company.

Crist, Darlene, Gail Scowcroft, and James Harding (2009). *World Ocean Census: A Global Survey of Marine Life*. Richmond Hill, Canada: Firefly Books.

Dinwiddie, Robert, Philip Eales, Sue Scott, et al. (2008). *Ocean: The World's Last Wilderness Revealed.* London: Dorling Kindersley.

Earle, Sylvia (2001). *National Geographic Atlas of the Ocean: The Deep Frontier*. Washington, D. C.: National Geographic.

Nouvian, Claire (2007). *The Deep: The Extraordinary Creatures of the Abyss*. Chicago, IL: University of Chicago Press.

Roberts, Callum (2007). *The Unnatural History of the Sea*. Washington, D.C.: Island Press.

Williams, Glyn (2013). *Naturalists at Sea: Scientific Travellers from Dampier to Darwin*. New Haven, CT: Yale University Press.

1.

2.

Author: Ernst Haeckel (1834–1919)
Title: *Kunstformen der Natur (Art forms of nature)*
Imprint: Leipzig:Verlag des Bibliographischen Instituts, 1899–1904

1. This stunning plate is taken from Ernst Haeckel's influential *Kunstformen der Natur* and features numerous sea anemones arranged in a naturalistic composition, displaying them, as in life, attached to rocks and with their feeding tentacles extended.

Author: George C. Perry
Title: *Arcana, or the museum of natural history, containing the most recent discovered objects, embellished with coloured plates and corresponding descriptions, with extracts relating to animals, and remarks of celebrated travellers: combining a general survey of nature. v. 1.*
Publisher: London: Printed by George Smeeton, 1811

2. From George Perry's *Arcana*, this seahorse is oddly contorted with its prehensile tail directed backward—a posture never seen in life. The curious, hairy mane along the animal's back is also rather fanciful. Nonetheless, the specimen is clearly a male, with a distended brood pouch probably concealing young inside.

INDEX